LOVE IS IN THE EARTH-

MINERALOGICAL PICTORIAL

Copyright © 1993 by
EARTH-LOVE PUBLISHING HOUSE
3440 Youngfield Street, Suite 353
Wheat Ridge, Colorado 80033 USA

Published by
EARTH-LOVE PUBLISHING HOUSE
3440 Youngfield Street, Suite 353
Wheat Ridge, Colorado, 80033 USA

First Printing 1994

This chemical compositions in this book are based upon "Dana's Textbook Of Mineralogy", Nickel & Nichols "Mineral Reference Manual", and "The Encyclopedia of Minerals and Gemstones".

Library of Congress Catalogue Card Number: 93-073772

ISBN: 0-9628190-2-6

Printed in Singapore

LOVE IS IN THE EARTH-

MINERALOGICAL PICTORIAL

Designed, Directed, and Arranged by ♪, *Melody* ♪♪

Primary Photography by Jim Hughes
Additional Photography by Melody, Len Cram, Franklin King, Dave Shrum, and David Benoit

Illustrated by Julianne Guilbault

Edited by Bob Jackson

EARTH-LOVE PUBLISHING HOUSE

3440 Youngfield Street, Suite 353
Wheat Ridge, CO. 80033

Cover Art: "**THE WATERFALL**" created by the mineral kingdom in the lovely land of Brazil.

"The Waterfall" is comprised of a flowing calcite druse, filled with sparkling quartz, and presenting the peaceful calm and the sacred energies of the universe to all.

Illustrations created by Julianne Guilbault.

I would like to thank Julianne Guilbault for creating the illustrations shown within **LOVE IS IN THE EARTH - MINERALOGICAL PICTORIAL**. Julianne and I have been friends through many lifetimes and have worked together toward the furtherance of the "brotherhood" of "All That Is" and toward the actualization of the inner light. She has been active in the crystal awareness movement and the practice of Native American spirituality for years and is truly the personification of creativity, both living and being the essence of originality and ingenuity. She has been involved in graphics design and illustrating for over twenty years, utilizing the mediums of watercolour, pastels, charcoal, pen and ink, and acrylic. I thank her also for her love, her encouragement, in all areas of my life and for and her support in the illustration of **"LOVE IS IN THE EARTH - MINERALOGICAL PICTORIAL"** and in the illustration of both **"LOVE IS IN THE EARTH - A KALEIDOSCOPE OF CRYSTALS"** and **"LOVE IS IN THE EARTH - LAYING-ON-OF-STONES"**. Julianne may be contacted c/o Earth-Love Publishing House.

Primary Photography by Jim Hughes

I would like to thank Jim Hughes for the quality time we shared during the photographing of the minerals in this book. Although Jim and I are relatively new friends, the rapport between us was established quickly and we both truly enjoyed the photography sessions. He always has a smile and a refreshing, gracious, and relaxing attitude. Jim's photography techniques bring the soul of beauty and the ever present essence of each subject to allow the panorama of energies to be shown. Jim's studio, "Footprints Fine Photography", is located in Aurora, Colorado, USA.

I am especially grateful and give my warmest thanks to Lynn Fielding, attorney, through whom this book and the preceding "Love Is In The Earth ....." books have become a reality. Lynn and I are very good friends and I sincerely appreciate his help, his guidance, his love, and his encouragement in pursuing my path. His expertise and his expediency are always available.

Front cover Photography created by Jim Hughes, Aurora, Colorado, USA

Cover colour separations by Peter Hoyt, Pacific Scanning, Medford, OR, USA.

The author may be contacted c/o Earth-Love Publishing House.

DEDICATION

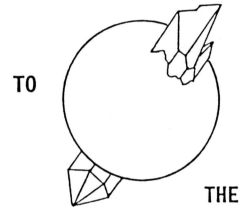

TO

THE

EARTH

ACKNOWLEDGEMENTS

I would like to thank the following people for their assistance in helping to make this book a reality.

Elenita, Natalino, Rackel, Gustavo, and Mariana Oliveira for my first introduction to the prime specimens of rutilated quartz and asterated quartz, for their true hospitality during the times we have shared, for their gifts of lovely mineral specimens, and for their love.

Jose Orizon, Ricardo, and Daniel de Almeida for my first introduction to the Faden Quartz from Brasil, for my first introduction to the zoned smokey/clear quartz configuration, for their true hospitality during the times we have shared, and for their love.

Bob Beck - for his being and for my first introduction to the Chevron amethyst.

Carlos do Prado Barbosa - for his being and for my first introduction to Phenacite.

Pai Carumbe - for his being and for our connection of understanding and love between Brasil and America.

Mike Brown - for his love and encouragement in all areas of my life, for his sensitivity to the sanctity of the Native American ways, and for his awareness and practice within crystal-consciousness.

Colorado Camera Co. - for their kindness and efficiency in providing for the development and associated requirements for the photographs within this book; service at Colorado Camera is always with a smile!

Edward Salisbury Dana - for teaching me, via his textbooks of mineralogy, about the multitude of crystalline structures which exist today.

Howard Dolph - for his being and his love; and for my first introduction to the Nephrite/Quartz structures, the violet sheen and electric-blue sheen Obsidian, and the lovely geode with square interior crystals which is shown within this book.

Monika Friedrich - for her love and encouragement, for her assistance in photographing in Madagascar, and for her lovely gifts of stones.

Arthur Goldstein - for my first introduction to Charoite from Russia.

Gypsy Blue Eyes - for his being and for his emanation of the ideal that minerals are truly alive.

Lucy and Joe Gross [Mama & Papa] - for their being, their love, encouragement, reassurance, motivation, patience, and support in all that I attempt and all that I am, for stimulating my interest in all of the kingdoms of the Earth, and for further helping me to both recognize and understand my heritage.

W. R. Horning - for his love and encouragement, for stimulating my interest in the mineral kingdom, and for my first introduction to many minerals.

Bob Jackson - for his love and encouragement in my life, for his patience, for his assistance in editing the text of the material, for making available [via Pleasant Company Ltd., Wheat Ridge, Colorado] the wonderful minerals of the world, for his creations as a silversmith/goldsmith, and for being a part of the many and varied adventures in our journey toward the self.

Marguerite Martin - in her memory and for stimulating my interest in the greater realm of minerals, and for her love and encouragement toward continuing interest in the mineral kingdom.

♫

Anthony Malakou - for his love and encouragement, for his efforts with Zimbabwe phenacite, and for my first introduction to the Ajoite phantom crystal from Africa.

Bruce and Barbara McDougall - for their love and encouragement, for my first introduction to Boulder Matrix Fire Opal, for their gifts of love, and for their lovely hospitality.

Max Gros Moyano - for teaching me about the alaranjado tourmaline and the ametista rutilado, and for his true hospitality during the time we shared.

Scot Nelson - for his love and encouragement.

Osho - for his being, in a time when many of us were ready to consciously experience.

Chris Pittario - for her unconditional love and total encouragement in all areas of my life.

Antar Pushkara - for his love and encouragement in my life, for being my kinsman, and for sharing his radiance and manifestation of inner calm, peace, and understanding.

Gregory Sluszka - for his love and encouragement, for giving of himself to further the advancement of all, and for his continuing support and assistance in the application of the mineral kingdom to elevate both collective and personal development.

Milt Szulinski - in his memory, and for his love and encouragement toward personal development.

Jean and Don Schroeder - for their love and encouragement.

LaSonda Sioux Sipe - for her continuing friendship, love, and understanding.

Rob Smith [African Gems & Minerals, Johannesburg, RSA] - for his love and encouragement, for my first introduction to the Ajoite crystal, for his lovely hospitality, for organizing and creating, and for making available the wonderful minerals from the African continent.

Layton Talbott - for his love and encouragement and for my first introduction to the gold/silver inclusions in quartz.

Angel Torrecillas - for my first introduction to Peacock Rock and to Ruby Silver, for his many gifts of minerals, and for his lovely smile.

Josef Vajdak - for his love and encouragement and for making the fine specimens of the world available to all through Pequa Rare Minerals, Massapequa, New York.

Liza Van De Linde - for her love and encouragement, for her openness and care, and for her friendship.

John Wittman - for his love and encouragement and for his true friendship.

In addition, I want thank all of those who shared their minerals for photographing; the credits are shown with each photograph. Thanks also to the Colorado School of Mines for allowing some of their mineral collection to be a part of this book. Thanks also to Forrest and Barbara Cureton (Cureton Mineral Co., Tucson, Az), Jim McGlasson (The Collector's Stope, Littleton, Colorado), David Shannon Minerals (Mesa, Arizona), for making available many of these mineral specimens to collectors throughout the world.

Thanks also to those friends who provided the quotations and poetry which are included herein. I especially want to thank all of those people throughout the world who have touched my life and have,

♫

To The Reader

This mineralogical pictorial has been published in order to provide photographs of the minerals discussed in "Love Is In The Earth - A Kaleidoscope Of Crystals" and "Love Is In The Earth - Laying-On-Of-Stones". The minerals within this book are typical specimens - they are provided in order to allow the reader to recognize the mineral, and/or to gain a visual understanding of a specific mineral. These minerals are those which are lovely for collectors and which are totally representative of the compositions discussed in prior books.

The mineralogical information accompanying the photographs has been obtained from "Dana's Textbook Of Mineralogy", Nickel & Nichols "Mineral Reference Manual", "The Encyclopedia of Minerals and Gemstones", and from mineralogists throughout the world.

The journey into the multi-faceted energies of the mineral kingdom has begun; the journey has continued toward the actualization of the perfect state. And finally, the reader of "Love Is In The Earth ..." has available actual photographic examples of the minerals previously discussed. It should be noted that there are additional minerals presented in this book - both for their beauty and for their lovely energies which have been experienced and documented during the last three years. May you continue to your state of fulfillment, always knowing that you are the wonder of the world. I wish you peace within yourself, love to guide you, and the understanding leading to bliss. May you know and experience love - from within, from upon, and from surrounding the Earth ♥

This book is arranged in alphabetical order to promote ease in locating photographs of minerals and of specific configurations. The listing of the contents follows and can be used as an index. The bold numbers indicate that the location of the photograph of the mineral is on the specified page; additional numbers listed, indicate another page on which the mineral is mentioned.

TABLE OF CONTENTS/INDEX

LOVE IS IN THE EARTH

♥

EARTH NOTES

EARTH NOTES

ACANTHITE - *Iron black crystalline; Composition: Ag_2S; Hardness 1-1.5; Locality: Joachimstal, Bohemia*

Photography by Jim Hughes, Assisted by ♪ Melody ♫
Collection of ♪ Melody ♫, Applewood, CO, USA

ACTINOLITE - *Greyish-green, Vitreous; Composition: $Ca_2(Mg,Fe)_5Si_8O_{22}(OH)_2$: Hardness 5-6; Locality: California, USA.*

Photography by Jim Hughes, Assisted by ♪ Melody ♫
Collection of Bob Jackson, Applewood, CO, USA

ACTINOLITE - *In Quartz; Composition: $Ca_2(Mg,Fe)_5Si_8O_{22}(OH)_2$; Hardness 5-6; Locality: Minas Gerais, Brasil*

Photography by Jim Hughes, Assisted by ♪ Melody ♫
Collection of ♪ Melody ♫, Applewood, CO, USA

ADAMITE - Vitreous Green:
Composition: $Zn_2AsO_4(OH)$;
Hardness 3.5; Locality: Mexico

Photography by Jim Hughes, Assisted by ♪ Melody ♫
Collection of ♪ Melody ♫, Applewood, CO, USA

ADULARIA: Nearly pure
potassium silicate; transparent to
translucent; Composition:
$KAlSi_3O_8$; Hardness 6; Locality:
Switzerland

Photography by Jim Hughes, Assisted by ♪ Melody ♫
Collection of ♪ Melody ♫, Applewood, CO, USA

AGATE - (Achat) - SiO_2 with
other polymorphs of silica -
Lustrous with transparency/
translucency; Hardness 7;
Locality: Germany;

Photography by Jim Hughes, Assisted by ♪ Melody ♫
Collection of ♪ Melody ♫, Applewood, CO, USA
Gift of W.R. Horning, California, USA

AGATE - (Angel Wing) - SiO₂ with other polymorphs of silica - Lustrous with transparency/ translucency; Hardness 7; Locality: Oregon, USA

Photography by Jim Hughes, Assisted by ♪ Melody ♫
Collection of Bob Jackson, Applewood, CO, USA

AGATE - (Angel Wing) - SiO₂ *with other polymorphs of silica -* *Lustrous with transparency/* *translucency; Hardness 7;* *Locality: Oregon, USA*

Photography by Jim Hughes, Assisted by ♪ Melody ♫
Collection of Bob Jackson, Applewood, CO, USA

AGATE - (Botryoidal Black) - *SiO₂ with other polymorphs of* *silica - Lustrous with* *transparency/translucency;* *Hardness 7; Locality; Idaho,* *USA*

Photography by Jim Hughes, Assisted by ♪ Melody ♫
Collection of Bob Jackson, Applewood, CO, USA

AGATE - (Black - Slabs) - SiO₂
with other polymorphs of silica -
Lustrous with transparency/
translucency; Hardness 7;
Locality: Minas Gerais, Brasil

Photography by Jim Hughes, Assisted by ♪ Melody ♫
Collection of ♪ Melody ♫ Applewood, CO, USA

AGATE - (Blue Lace) - SiO₂ with
other polymorphs of silica -
Lustrous with transparency/
translucency; Hardness 7;
Locality: Republic of South
Africa, Africa

Photography by Jim Hughes, Assisted by ♪ Melody ♫
Collection of ♪ Melody ♫, Applewood, CO, USA

AGATE - (Botswana) SiO₂ with
other polymorphs of silica -
Lustrous with transparency/
translucency; Hardness 7:
Locality: Botswana, Africa

Photography by Jim Hughes, Assisted by ♪ Melody ♫
Collection of Bob Jackson, Applewood, CO, USA

AGATE - (Brasilian) - SiO₂ with other polymorphs of silica - Lustrous with transparency/ translucency; Hardness 7; Locality: Rio Grande Do Sul, Brasil

Photography by Jim Hughes, Assisted by ♪ Melody ♫
Collection of ♪ Melody ♫, Applewood, CO, USA

AGATE - (Dendritic) - SiO₂ with other polymorphs of silica - Lustrous with transparency/ translucency; Hardness 7; Locality: Mexico

Photography by Jim Hughes, Assisted by ♪ Melody ♫
Collection of ♪ Melody ♫, Applewood, CO, USA

AGATE - (Dry Head) - SiO₂ with other polymorphs of silica - Lustrous with transparency/ translucency; Hardness 7; Locality: Oregon

Photography by Jim Hughes, Assisted by ♪ Melody ♫
Collection of ♪ Melody ♫, Applewood, CO, USA
Gift from Bob Jackson, Applewood, CO, USA

AGATE - (Ellensburg Blue) - SiO₂ with other polymorphs of silica - Lustrous with transparency/translucency; Hardness 7; Locality: Washington, USA

Photography by Jim Hughes, Assisted by ♪ Melody ♫
Collection of ♪ Melody ♫, Applewood, CO, USA

AGATE - (Fire) - SiO₂ with other polymorphs of silica - Lustrous with transparency/translucency; Hardness 7; Locality: Mexico

Photography by Jim Hughes, Assisted by ♪ Melody ♫
Collection of Bob Jackson, Applewood, CO, USA

AGATE - (Flame) - SiO₂ with other polymorphs of silica - Lustrous with transparency/ translucency; Hardness 7; Locality: Mexico

Photography by Jim Hughes, Assisted by ♪ Melody ♫
Collection of Bob Jackson, Applewood, CO, USA

AGATE - (Holly Blue) - SiO$_2$ with other polymorphs of silica - Lustrous with transparency/translucency; Hardness 7; Locality: Oregon, USA
Photography by Jim Hughes, Assisted by ♪ Melody ♫
Collection of ♪ Melody ♫, Applewood, CO, USA

AGATE - (Holly Blue) - SiO$_2$ with other polymorphs of silica - Lustrous with transparency/translucency; Hardness 7; Locality: Oregon, USA
Photography by Jim Hughes, Assisted by ♪ Melody ♫
Collection of ♪ Melody ♫, Applewood, CO, USA

AGATE - (Iris) - SiO₂ with other polymorphs of silica - Lustrous with transparency/translucency; Hardness 7; Locality: Oregon, USA

Photography by Jim Hughes, Assisted by ♪ Melody ♫
Collection of Bob Jackson, Applewood, CO, USA

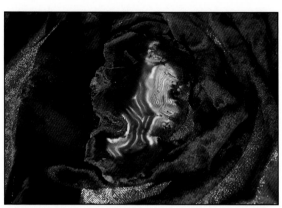

AGATE - (Laguna) - SiO₂ with other polymorphs of silica - Lustrous with transparency/ translucency; Hardness 7; Locality: Mexico

Photography by Jim Hughes, Assisted by ♪ Melody ♫
Collection of Bob Jackson, Applewood, CO, USA

AGATE - (Lake Superior) - SiO₂ with other polymorphs of silica - Lustrous with transparency/ translucency; Hardness 7; Locality: Michigan, USA

Photography by Jim Hughes, Assisted by ♪ Melody ♫
Collection of Bob Jackson, Applewood, CO, USA

AGATE - (Mexican Lace) - SiO$_2$ with other polymorphs of silica - Lustrous with transparency/ translucency; Hardness 7; Locality: Mexico

Photography by Jim Hughes, Assisted by ♪ Melody ♫
Collection of Bob Jackson, Applewood, CO, USA

AGATE - (Montana) - SiO$_2$ with other polymorphs of silica - Lustrous with transparency/ translucency; Hardness 7; Locality Montana, USA

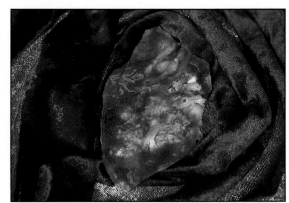

Photography by Jim Hughes, Assisted by ♪ Melody ♫
Collection of Bob Jackson, Applewood, CO, USA

AGATE - (Moss) - SiO$_2$ with other polymorphs of silica - Lustrous with transparency/ translucency; Hardness 7; Locality: Oregon, USA

Photography by Jim Hughes, Assisted by ♪ Melody ♫
Collection of Bob Jackson, Applewood, CO, USA

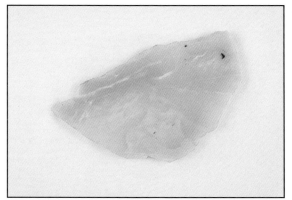

AGATE - (Ocean Spray) - SiO$_2$ with other polymorphs of silica - Lustrous with transparency/ translucency; Hardness 7; Locality: Canada

Photography by Jim Hughes, Assisted by ♪ Melody ♫
Collection of Bob Jackson, Applewood, CO, USA

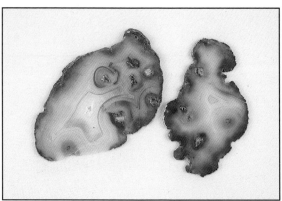

AGATE - (Orbicular Black) - SiO$_2$ with other polymorphs of silica - Lustrous with transparency/ translucency; Hardness 7; Locality: Idaho, USA

Photography by Jim Hughes, Assisted by ♪ Melody ♫
Collection of Bob Jackson, Applewood, CO, USA

AGATE - (Plasma) - SiO$_2$ with other polymorphs of silica - Lustrous with transparency/ translucency; Hardness 7; Locality: Oregon, USA

Photography by Jim Hughes, Assisted by ♪ Melody ♫
Collection of ♪ Melody ♫, Applewood, CO, USA

AGATE - (Priday Plume) - SiO$_2$
with other polymorphs of silica -
Lustrous with transparency/
translucency; Hardness 7;
Locality: Oregon, USA

Photography by Jim Hughes, Assisted by ♪ Melody ♫
Collection of ♪ Melody ♫, Applewood, CO, USA

AGATE - (Priday Plume) - SiO$_2$
with other polymorphs of silica -
Lustrous with transparency/
translucency; Hardness 7;
Locality: Oregon, USA

Photography by Jim Hughes, Assisted by ♪ Melody ♫
Collection of Bob Jackson, Applewood, CO, USA

AGATE - (Regency Rose) - SiO$_2$
with other polymorphs of silica -
Lustrous with transparency/
translucency; Hardness 7;
Locality: Oregon, USA

Photography by Jim Hughes, Assisted by ♪ Melody ♫
Collection of Bob Jackson, Applewood, CO, USA

AGATE - (Rose-Eye) - SiO₂ with other polymorphs of silica - Lustrous with transparency/translucency; Hardness 7; Locality: Mexico

Photography by Jim Hughes, Assisted by ♪ Melody ♫
Collection of ♪ Melody ♫, Applewood, CO, USA

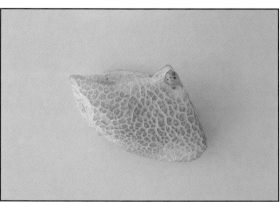

AGATE - (Snakeskin) - SiO₂ with other polymorphs of silica - Lustrous with transparency/translucency; Hardness 7; Locality: India

Photography by Jim Hughes, Assisted by ♪ Melody ♫
Collection of ♪ Melody ♫, Applewood, CO, USA

AGATE - (Stinking Water Plume) SiO₂ with other polymorphs of silica - Lustrous with transparency/translucency; Hardness 7; Locality: Oregon, USA

Photography by Jim Hughes, Assisted by ♪ Melody ♫
Collection of Bob Jackson, Applewood, CO, USA

AGATE - (Turritella) - SiO₂
with other polymorphs of silica -
Lustrous with transparency/
translucency; Hardness 7;
Locality: Minnesota, USA

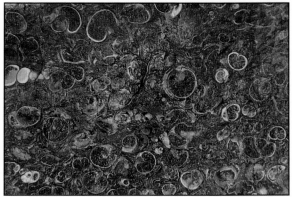

Photography by Jim Hughes, Assisted by ♪ Melody ♫
Collection of ♪ Melody ♫, Applewood, CO, USA

AGATE - (Woodward Ranch
Plume) - SiO₂ with other
polymorphs of silica - Lustrous
with transparency/translucency;
Hardness 7; Locality: Texas,
USA

Photography by Jim Hughes, Assisted by ♪ Melody ♫
Collection of Bob Jackson, Applewood, CO, USA

AJOITE - Quartz with
turquoise-coloured inclusions;
Quartz hardness 7;
Composition:
$(K,Na)Cu_7^{+2}AlSi_9O_{24}(OH)_6$♥$3H_2O$
Messina, Republic Of South
Africa, Africa

Photography by Jim Hughes, Assisted by ♪ Melody ♫
Collection of ♪ Melody ♫, Applewood, CO, USA

AJOITE - *Quartz with turquoise-coloured inclusions; Quartz hardness 7; Composition: $(K,Na)Cu_7^{+2}AlSi_9O_{24}(OH)_6 \heartsuit 3H_2O$ Messina, Republic Of South Africa, Africa*

Photography by Jim Hughes, Assisted by ♪ Melody ♫
Collection of ♪ Melody ♫, Applewood, CO, USA

ALABASTER - *Fine grained often impure with clay, calcium carbonate or silica; Composition: $CaSO_4 \heartsuit 2H_2O$; Hardness 1.5-2; Locality: Arizona, USA*

Photography by Jim Hughes, Assisted by ♪ Melody ♫
Collection of Bob Jackson, Applewood, CO, USA

ALABASTER - *Fine grained often impure with clay, calcium carbonate or silica; Composition: $CaSO_4 \heartsuit 2H_2O$; Hardness 1.5-2; Locality: Arizona, USA*

Photography by Jim Hughes, Assisted by ♪ Melody ♫
Collection of Bob Jackson, Applewood, CO, USA

ALBITE - *(With Chlorite in/on Quartz)* - *Vitreous/pearly white, bluish, grey, pink, etc. (Albite); Composition: $NaAlSi_3O_8$ with chlorite and quartz; Hardness (Albite) 6-6.5; Locality: Messina, Republic of South Africa, Africa*

Photography by Jim Hughes, Assisted by ♪ Melody ♫
Collection of ♪ Melody ♫, Applewood, CO, USA

ALEXANDRITE - - *Emerald green, columbine-red by artificial light, crystals transparent to translucent with appearance of brown until held to light; Composition: $BeAl_2O_{3-4}$; Hardness 8.5; Locality: Zimbabwe, Africa*

Photography by Jim Hughes, Assisted by ♪ Melody ♫
Collection of ♪ Melody ♫, Applewood, CO, USA

ALMANDINE - *(In Quartz); Vitreous/resinous red/brown/black Almadine; Composition: $Fe_3Al_2(SiO_4)_3$ with Quartz; Hardness (Almadine) 6.5-7.5; Locality: Pakistan*

Photography by Jim Hughes, Assisted by ♪ Melody ♫
Collection of ♪ Melody ♫, Applewood, CO, USA

ALMANDINE - *(Black)* - *Vitreous/resinous; Composition:* $Fe_3Al_2(SiO_4)_3$; *Locality: Canada*

Photography by Jim Hughes, Assisted by ♪ Melody ♫
Collection of Bob Jackson, Applewood, CO, USA

ALMANDINE - *(In Quartz);* $Fe_3Al_2(SiO_4)_3$ *Vitreous/resinous red/brown/black Almadine; Composition:* $Fe_3Al_2(SiO_4)_3$ *(Almadine); Hardness (Almadine) 6.5-7.5; Locality: Pakistan*

Photography by Jim Hughes, Assisted by ♪ Melody ♫
Collection of ♪ Melody ♫, Applewood, CO, USA

ALURGITE - *A copperish-red potassium-magnesium mica with small amounts of manganese; Hardness 2.5 - 3; Locality: Piedmont, Italy.*

Photography by Jim Hughes, Assisted by ♪ Melody ♫
Collection of ♪ Melody ♫, Applewood, CO, USA

AMAZONITE - Transparent to translucent blue-green; Composition: $KAlSi_3O_8$; Hardness 6-6.5; Colorado, USA.

AMAZONITE - (With Smokey Quartz) - Transparent to translucent blue-green; Composition: $KAlSi_3O_8$; Hardness 6-6.5 (Amazonite); Colorado, USA.

AMBER - Lustrous yellow, brown, yellowish-white; Composition: [C,H,O]; Hardness 2 - 2.5; Locality: Baltic Sea

AMBLYGONITE - *Vitreous/ lustrous white, yellowish, pink, greenish, etc.; Composition: $(Li,Na)AlPO_4(F,OH)$; Hardness 5.5 - 6; Locality: Connecticut, USA*

Photography by Jim Hughes, Assisted by ♪ Melody ♫
Collection of ♪ Melody ♫, Applewood, CO, USA

AMETHYST - *(Brandenberg) - SiO_2 with ferric iron - light purple to dark purple; Hardness 7; Locality: Republic of South Africa, Africa*

Photography by Jim Hughes, Assisted by ♪ Melody ♫
Collection of ♪ Melody ♫, Applewood, CO, USA

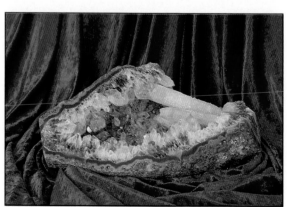

AMETHYST - *(With Calcite and Clear Quartz) - SiO_2 with ferric iron - light purple to dark purple; Hardness 7; Locality: Minas Gerais, Brasil*

Photography by Jim Hughes, Assisted by ♪ Melody ♫
Collection of ♪ Melody ♫, Applewood, CO, USA

AMETHYST - (Chevron) - SiO₂
with ferric iron - light purple to dark purple; Hardness 7; Locality: India

Photography by Jim Hughes, Assisted by ♪ Melody ♫
Collection of Bob Jackson, Applewood, CO, USA

AMETHYST - (Chevron) - SiO₂
with ferric iron - light purple to dark purple; Hardness 7; Locality: India

Photography by Jim Hughes, Assisted by ♪ Melody ♫
Collection of ♪ Melody ♫, Applewood, CO, USA

AMETHYST -
(With Cacoxenite) - SiO₂ with ferric iron - light purple to dark purple, and Cacoxenite; Hardness 7 (Amethyst); Locality: Minas Gerais, Brasil

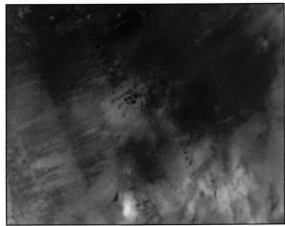

Photography by Jim Hughes, Assisted by ♪ Melody ♫
Collection of Bob Jackson, Applewood, CO, USA

AMETHYST - (With Cacoxenite)
SiO_2 *with ferric iron - light purple to dark purple, and Cacoxenite; Hardness 7 (Amethyst); Locality: Minas Gerais, Brasil*

Photography by Jim Hughes, Assisted by ♪ Melody ♫
Collection of Bob Jackson, Applewood, CO, USA

AMETHYST - With Chlorite - Serifos, Greece
Photography by Jim Hughes, Assisted by ♪ Melody ♫
Collection of Gregory Sluszka, USA

AMETHYST - (Drusy) - SiO$_2$ with ferric iron - light purple to dark purple; Hardness 7; Locality: Republic of South Africa, Africa

Photography by Jim Hughes, Assisted by ♪ Melody ♫
Collection of Bob Jackson, Applewood, CO, USA

AMETHYST - (Elestiated Quartz) - SiO$_2$ with ferric iron - light purple to dark purple; Hardness 7; Locality: Minas Gerais, Brasil

Photography by Jim Hughes, Assisted by ♪ Melody ♫
Collection of ♪ Melody ♫, Applewood, CO, USA

AMETHYST - (With Fluorite) - SiO$_2$ with ferric iron - light purple to dark purple; Hardness 7; Locality: Arizona, USA

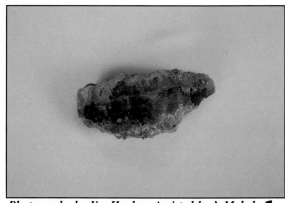

Photography by Jim Hughes, Assisted by ♪ Melody ♫
Collection of Bob Jackson, Applewood, CO, USA

AMETHYST - (Oregon) - SiO$_2$ with small amount of ferric iron - very light purple; Hardness 7; Locality: Oregon, USA

Photography by Jim Hughes, Assisted by ♪ Melody ♫
Collection of ♪ Melody ♫, Applewood, CO, USA

AMETHYST - (Phantom) - SiO$_2$ with ferric iron - light purple to dark purple; Hardness 7; Locality: Namibia, Africa

Photography by Jim Hughes, Assisted by ♪ Melody ♫
Collection of ♪ Melody ♫, Applewood, CO, USA

AMETHYST - (Stalactite)- SiO$_2$ with ferric iron - light purple to dark purple; Hardness 7; Locality: Peru

Photography by Jim Hughes, Assisted by ♪ Melody ♫
Collection of ♪ Melody ♫, Applewood, CO, USA

AMETHYST - (Stalactite) - SiO_2 with ferric iron - light purple to dark purple; Hardness 7; Locality: Mexico

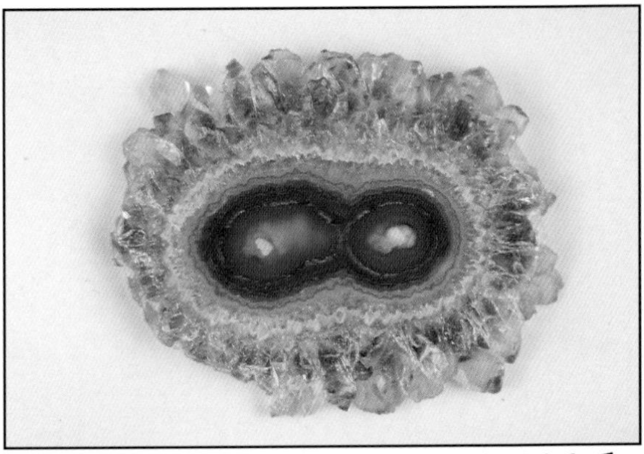

Photography by Jim Hughes, Assisted by ♪ Melody ♫
Collection of ♪ Melody ♫, Applewood, CO, USA

AMETHYST - (Vera Cruz) - SiO_2 with ferric iron - light to medium delicate purple to lavender; Hardness 7; Locality: Mexico

Photography by Jim Hughes, Assisted by ♪ Melody ♫
Collection of ♪ Melody ♫, Applewood, CO, USA

AMETRINE - (Elestial) - Combination of Amethyst and Citrine; Hardness 7; Locality: Bolivia, South America

Photography by Jim Hughes, Assisted by ♪ Melody ♫
Collection of ♪ Melody ♫, Applewood, CO, USA

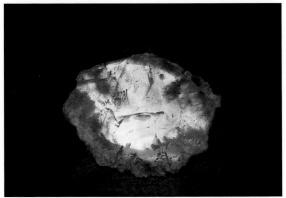

AMETRINE - (Slice) -
Combination of Amethyst and
Citrine; Hardness 7; Locality:
Bolivia, South America

Photography by Jim Hughes, Assisted by ♪ Melody ♫
Collection of Jackson & Melody, Applewood, CO, USA

AMETRINE - (Crystal) -
Combination of Amethyst and
Citrine; Hardness 7;
Locality:Bolivia, South America

Photography by Jim Hughes, Assisted by ♪ Melody ♫
Collection of ♪ Melody ♫, Applewood, CO, USA

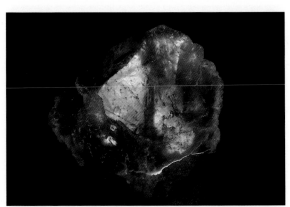

AMETRINE - (Slice) -
Combination of Amethyst and
Citrine; Hardness 7; Locality:
Bolivia, South America

Photography by Jim Hughes, Assisted by ♪ Melody ♫
Collection of Jackson and Melody Applewood, CO, USA

AMETRINE *- (Crystal) - Combination of Amethyst and Citrine; Hardness 7; Minas Gerais, Brasil*

Photography by Jim Hughes, Assisted by ♪ Melody ♫
Collection of ♪ Melody ♫, Applewood, CO, USA
Gift of Julianne Guilbault

AMMONITE *- Fossilized snail-like animal*

Photography by Jim Hughes, Assisted by ♪ Melody ♫
Collection of ♪ Melody ♫, Applewood, CO, USA

ANALCIME *- Vitreous white, colourless, grey, pink, greenish, and/or yellowish; Composition: $Na(AlSi_2)O_6$♥H_2O; Hardness 5 to 5.5; Locality: Colorado, USA*

Photography by Jim Hughes, Assisted by ♪ Melody ♫
Collection of ♪ Melody ♫, Applewood, CO, USA

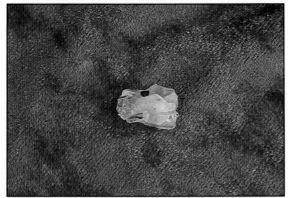

ANATASE (On Quartz) -
Adamantine brown, green, grey,
black, etc.; Composition: TiO_2;
Hardness 5.5 - 6; Locality:
France

Photography by Jim Hughes, Assisted by ♪ Melody ♫
Collection of ♪ Melody ♫, Applewood, CO, USA

ANDALUSITE - Translucent
pink, white red, grey, yellow,
violet, etc.; Composition: Al_2SiO_5
Hardness 6.5 - 7.5; Locality:
Andalusia, Spain

Photography by Jim Hughes, Assisted by ♪ Melody ♫
Collection of Bob Jackson, Applewood, CO, USA

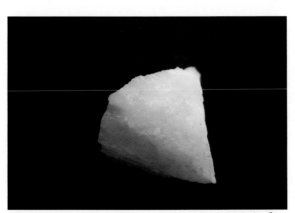

ANDESINE - (Oligoclase) -
Occurs in granular and volcanic
rocks of medium silica content;
Composition: $(nNaAlSi_3O_8 -$
$mCaAl_2Si_2O_{8)}$; Hardness 5 - 6;
Locality: North Carolina, USA

Photography by Jim Hughes, Assisted by ♪ Melody ♫
Collection of ♪ Melody ♫ Applewood, CO, USA

ANDRADITE - $Ca_3Fe_2(SiO_4)_3$ - *Vitreous/resinous greenish-yellow, green, brown, black, etc.; Hardness 6.5-7.5; Locality: Arizona, USA*

Photography by Jim Hughes, Assisted by ♪ Melody ♫
Collection of ♪ Melody ♫, Applewood, CO, USA

ANDRADITE - $Ca_3Fe_2(SiO_4)_3$ - *Vitreous/resinous greenish-yellow, green, brown, black, etc.; Hardness 6.5-7.5; Locality: Russia*

Photography by Jim Hughes, Assisted by ♪ Melody ♫
Collection of ♪ Melody ♫, Applewood, CO, USA
Gift of LaSonda Sioux Sipe

ANGELITE - $CaSO_4$ - *Vitreous/lustrous blue; Hardness 3.5; Locality: Peru, South America*

Photography by Jim Hughes, Assisted by ♪ Melody ♫
Collection of Bob Jackson, Applewood, CO, USA

ANGLESITE - PbSO₄ - Adamantine/resinous white, grey, yellow, green, blue; Hardness 2.5 to 3; Locality: Isle Of Anglesey, Wales

Photography by Jim Hughes, Assisted by ♪ Melody ♫
Collection of Bob Jackson, Applewood, CO, USA

ANHYDRITE - (Crystal) - CaSO₄ - Vitreous/lustrous colourless, bluish, violet; Hardness 3.5; Locality: Italy

Photography by Jim Hughes, Assisted by ♪ Melody ♫
Collection of ♪ Melody ♫, Applewood, CO, USA

ANTHOPHYLLITE - (Mg,Fe)₇Si₈O₂₂(OH)₂ - Vitreous brownish-grey, yellowish-brown, green; Hardness 5-5.6; Locality: Connecticut, USA

Photography by Jim Hughes, Assisted by ♪ Melody ♫
Collection of ♪ Melody ♫, Applewood, CO, USA

ANTIMONY - *Sb - Metallic white; Hardness 3-3.5; Locality: Quebec, Canada*

Photography by Jim Hughes, Assisted by ♪ Melody ♫
Collection of ♪ Melody ♫, Applewood, CO, USA

ANTLERITE - *(With Malachite and Azurite)-* $Cu_3SO_4(OH)_4$ - *Vitreous* <u>brown/green</u>; *Hardness 3.5; Locality: Arizona, USA*

Photography by Jim Hughes, Assisted by ♪ Melody ♫
Collection of ♪ Melody ♫, Applewood, CO, USA

ANYOLITE - $Ca_2Al_3(SiO_4)_3(OH)$ *with chromium; vitreous lustrous green; Hardness 6-6.5; Locality: Tanzania, Africa*

Photography by Jim Hughes, Assisted by ♪ Melody ♫
Collection of ♪ Melody ♫, Applewood, CO, USA

APACHE GOLD - *Combination of Steatite and Pyrite; Locality: Mexico*

Photography by Jim Hughes, Assisted by ♪ Melody ♫
Collection of ♪ Melody ♫, Applewood, CO, USA

APATITE - $Ca_5(PO_4)_3(F,OH,Cl)$ - *Vitreous, extensive colour range; (Yellow colour in photograph); Hardness 5; Locality: Mexico*

Photography by Jim Hughes, Assisted by ♪ Melody ♫
Collection of ♪ Melody ♫, Applewood, CO, USA

APATITE - $Ca_5(PO_4)_3(F,OH,Cl)$ - *Vitreous, extensive colour range; (Blue colour in photograph); Hardness 5; Locality: Minas Gerais, Brasil*

Photography by Jim Hughes, Assisted by ♪ Melody ♫
Collection of Bob Jackson, Applewood, CO, USA

*APATITE - Ca$_5$(PO$_4$)$_3$(F,OH,Cl)
- Vitreous, extensive colour
range; (Gold colour in
photograph); Hardness 5;
Locality: Mexico*

*Photography by Jim Hughes, Assisted by ♪ Melody ♫
Collection of ♪ Melody ♫, Applewood, CO, USA*

*APATITE - Ca$_5$(PO$_4$)$_3$(F,OH,Cl)
- Vitreous, extensive colour
range; (Purple colour in
photograph); Hardness 5;
Locality: Mexico*

*Photography by Jim Hughes, Assisted by ♪ Melody ♫
Collection of ♪ Melody ♫, Applewood, CO, USA*

*APOPHYLLITE -
KF ♥ Ca$_4$(SiO$_5$)$_4$ ♥ 8H$_2$O;
Transparent to opaque white,
greyish, <u>greenish</u>, yellowish,
rose-red tint; Hardness 4.5-5;
Locality: Pune, India*

*Photography by Jim Hughes, Assisted by ♪ Melody ♫
Collection of ♪ Melody ♫, Applewood, CO, USA*

APOPHYLLITE - (Pyramidal Structure) - KF♥Ca₄(SiO₅)₄♥8H₂O; *Transparent to opaque white*, greyish, greenish, yellowish, rose-red tint; Hardness 4.5-5; Locality: Pune, India

Photography by Jim Hughes, Assisted by ♪ Melody ♫
Collection of ♪ Melody ♫, Applewood, CO, USA

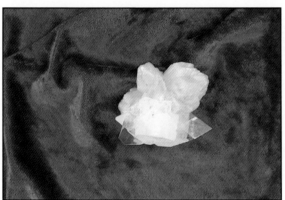

APOPHYLLITE - (In Cluster) - KF♥Ca₄(SiO₅)₄♥8H₂O; Transparent to opaque *white*, greyish, greenish, yellowish, rose-red tint; Hardness 4.5-5; Locality: Pune, India

Photography by Jim Hughes, Assisted by ♪ Melody ♫
Collection of ♪ Melody ♫, Applewood, CO, USA

APOPHYLLITE - (Cluster) - KF♥Ca₄(SiO₅)₄♥8H₂O; Transparent to opaque white, greyish, greenish, yellowish, rose-red tint; Hardness 4.5-5; Locality: Pune, India

Photography by Jim Hughes, Assisted by ♪ Melody ♫
Collection of ♪ Melody ♫, Applewood, CO, USA

AQUAMARINE - (Crystal) $Be_3Al_2(SiO_3)_6$ - Vitreous/resinous blue to bluish-green;
Hardness 7.5-8; Locality: Minas Gerais, Brasil.
Photography by Jim Hughes, Assisted by ♪ Melody ♫
Collection of ♪ Melody ♫, Applewood, CO, USA

AQUAMARINE - (Crystal)
$Be_3Al_2(SiO_3)_6$ - Vitreous/resinous
blue to bluish-green; Hardness
7.5-8; Locality: Afghanistan

Photography by Jim Hughes, Assisted by ♪ Melody ♫
Collection of ♪ Melody ♫, Applewood, CO, USA

ARAGONITE - $CaCO_3$ - Vitreous
colourless white, grey, yellowish,
blue, etc.; Hardness 3.5-4;
Locality: Aragon, Spain.

Photography by Jim Hughes, Assisted by ♪ Melody ♫
Collection of ♪ Melody ♫, Applewood, CO, USA

ARAGONITE - $CaCO_3$ - Vitreous
colourless white, grey, yellowish,
blue, etc.; Hardness 3.5-4;
Locality: Mexico

Photography by Jim Hughes, Assisted by ♪ Melody ♫
Collection of ♪ Melody ♫, Applewood, CO, USA

ARAGONITE - *CaCO₃* - *Vitreous colourless white, grey, yellowish, blue, etc.; Hardness 3.5-4; Locality: Mexico*

Photography by Jim Hughes, Assisted by ♪ Melody ♫
Collection of ♪ Melody ♫, Applewood, CO, USA

ARTHURITE - *(With Pharmocosiderite)* - *CuFe₂(AsO₄)₂(OH)₂♥4H₂O* - *Translucent pale olive green; Hardness 2.5; Locality: Cornwall, England*

Photography by Jim Hughes, Assisted by ♪ Melody ♫
Collection of ♪ Melody ♫, Applewood, CO, USA

ARTINITE - *Mg₂CO₃(OH)₂♥3H₂O - Vitreous satiny white; Hardness 2.5; Locality: Lombardia, Italy*

Photography by Jim Hughes, Assisted by ♪ Melody ♫
Collection of ♪ Melody ♫, Applewood, CO, USA

ASTROPHYLLITE -
$(K,Na)_3(Fe,Mn)_7Ti_2Si_8(O,OH)_{31}$ -
Sub-metallic/pearly bronze-yellow,
golden-yellow; Hardness 3;
Locality: Colorado, USA

Photography by Jim Hughes, Assisted by ♪ Melody ♫
Collection of ♪ Melody ♫, Applewood, CO, USA

ATELESTITE - $Bi_2OAsO_4(OH)$ -
Resinous/adamantine yellow;
Hardness 4.5-5; Saxony, Germany

Photography by Jim Hughes, Assisted by ♪ Melody ♫
Collection of Colorado School of Mines, Colorado, USA

AUGELITE - $Al_2PO_4(OH)_3$ -
Vitreous colourless, white,
yellowish, pale rose; Hardness
4.5-5; Locality: Skane, Sweden

Photography by Jim Hughes, Assisted by ♪ Melody ♫
Collection of ♪ Melody ♫, Applewood, CO, USA

AUGELITE - $Al_2PO_4(OH)_3$ - *Vitreous colourless, white, yellowish, pale rose; Hardness 4.5-5; Skane, Sweden*

Photography by Jim Hughes, Assisted by ♪ Melody ♫
Collection of ♪ Melody ♫, Applewood, CO, USA

AUGITE - $(Ca,Mg,Fe)_2(Si,Al)_2O_6$ - *Vitreous resinous brown, green, black; Hardness 5.5-6; Locality, Arizona, USA*

Photography by Jim Hughes, Assisted by ♪ Melody ♫
Collection of ♪ Melody ♫, Applewood, CO, USA

AURICHALCITE - $(Zn,Cu)_5(CO_3)_2(OH)_6$ - *Silky pearly pale green, greenish-blue, blue; Hardness 1-2; Locality: Mexico*

Photography by Jim Hughes, Assisted by ♪ Melody ♫
Collection of ♪ Melody ♫, Applewood, CO, USA

AUTENITE -
$Ca(UO_2)_2(PO_4)_2 \cdot 10H_2O$ - *Vitreous pearly yellow, greenish-yellow, pale green; Hardness 2-2.5; Locality: Saone-et-Loire, France*

Photography by Jim Hughes, Assisted by ♪ Melody ♫
Collection of ♪ Melody ♪, Applewood, CO, USA

AVENTURINE - *SiO_2 with impurities of mica, hematite, etc.; Vitreous lustrous green, etc.; Hardness 7: Locality: India*

Photography by Jim Hughes, Assisted by ♪ Melody ♫
Collection of ♪ Melody ♫, Applewood, CO, USA

AVOGADRITE - *$(K,Cs)BF_4$ - Translucent colourless, white yellowish, reddish; Hardness 2.5; Locality: Campania, Italy*

Photography by Jim Hughes, Assisted by ♪ Melody ♫
Collection of ♪ Melody ♫, Applewood, CO, USA

AXINITE -
$(Ca,Fe,Mg,Mn)_3Al_2BSi_4O_{15}(OH)$
*[General borosilicate]; Hardness
6-7; Locality: California, USA*

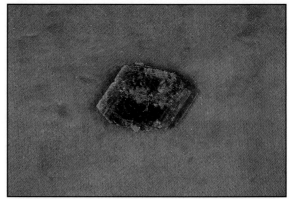

*Photography by Jim Hughes, Assisted by ♪ Melody ♫
Collection of ♪ Melody ♫, Applewood, CO, USA*

AZEZTULITE - *Composition
not defined; Vitreous colourless
to white; Hardness not yet
defined: Locality USA*

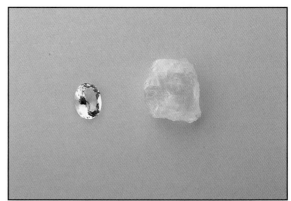

*Photograph courtesy of Bob Simmons and Kathy Warner,
Heaven & Earth, Vermont, USA; Photograph by David
Benoit, Massachusetts, USA*

AZULICITE - *($KAlSi_3O_8$) coated
with Chlorite and Limonite;
Colour yellow with blue
adularescence; Hardness 6;
Locality: Mexico*

*Photography by Jim Hughes, Assisted by ♪ Melody ♫
Collection of ♪ Melody ♫, Applewood, CO, USA*

AZURITE - Cu₃(CO₃)₂(OH)₂ - Vitreous blue; Hardness 3.5-4;
Locality: Tsumeb, Namibia
Photography by Jim Hughes, Assisted by ♪ Melody ♫
Collection of Bob Jackson, Applewood, CO, USA

AZURITE - Cu₃(CO₃)₂(OH)₂ - Vitreous blue crystals; Hardness 3.5-4;
Locality: Tsumeb, Namibia
Photography by Jim Hughes, Assisted by ♪ Melody ♫
Collection of ♪ Melody ♫, Applewood, CO, USA

AZURITE *- $Cu_3(CO_3)_2(OH)_2$ - Vitreous blue (Gemmy); Hardness 3.5-4; Locality: Arizona, USA*

Photography by Jim Hughes, Assisted by ♪ Melody ♫
Collection of ♪ Melody ♫, Applewood, CO, USA

AZURITE-MALACHITE *(Nodule) - Combination of Azurite and Malachite; Locality: Arizona, USA*

Photography by Jim Hughes, Assisted by ♪ Melody ♫
Collection of ♪ Melody ♫, Applewood, CO, USA

AZURITE-CHRYSOCOLLA-MALACHITE *- Combination; Locality: Republic of South Africa*

Photography by Jim Hughes, Assisted by ♪ Melody ♫
Collection of ♪ Melody ♫, Applewood, CO, USA

BABINGTONITE -
$CA_2(Fe,Mn)FeSi_5O_{14}(OH)$ -
Vitreous dark greenish-black;
Hardness 5.5-6; Locality:
Arendal, Norway.

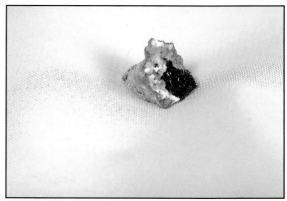

Photography by Jim Hughes, Assisted by ♪ Melody ♫
Collection of ♪ Melody ♫, Applewood, CO, USA

BACULITES - A genus of
polythalamous or many-
chambered cephalopods (only
known in the fossil state); Listed
here due to their less known
existence.

Photography by Jim Hughes, Assisted by ♪ Melody ♫
Collection of ♪ Melody ♫, Applewood, CO, USA

BARITE - (Crystal) $BaSO_4$ -
Vitreous resinous colourless,
blue, white, yellow, brown, grey,
etc.; Hardness 3-3.5; Locality:
Colorado, USA

Photography by Jim Hughes, Assisted by ♪ Melody ♫
Collection of Bob Jackson, Applewood, CO, USA

BARITE - (Oak) - BaSO$_4$ -
Vitreous resinous colourless, blue,
white, yellow, <u>brown,</u> grey, etc.;
Hardness 3-3.5; Locality: England

Photography by Jim Hughes, Assisted by ♪ Melody ♫
Collection of Bob Jackson, Applewood, CO, USA

BARITE - BaSO$_4$ - Vitreous
resinous colourless, blue, white,
yellow, brown, grey, etc.;
Hardness 3-3.5; Locality:
California, USA

Photography by Jim Hughes, Assisted by ♪ Melody ♫
Collection of Bob Jackson, Applewood, CO, USA

BARITE - (Rose) BaSO$_4$ -
Vitreous resinous colourless, blue,
white, yellow, brown, grey, etc.;
Hardness 3-3.5; Locality:
Oklahoma, USA

Photography by Jim Hughes, Assisted by ♪ Melody ♫
Collection of ♪ Melody ♫, Applewood, CO, USA

BASALT - *Dense fine-grained igneous in dark colours brown, red, green, black; Comprised of a soda-lime feldspar with pyroxene, iron ore, biotite, hornblende, etc.; Hardness varies with impurities; Locality: Washington, USA*

Photography by Jim Hughes, Assisted by ♪ Melody ♫
Collection of Bob Jackson, Applewood, CO, USA

BASINITE - *Very fine-grained Basalt; Locality: Idaho*

Photography by Jim Hughes, Assisted by ♪ Melody ♫
Collection of Bob Jackson, Applewood, CO, USA

BAVENITE - $Ca_4(Al,Be)_4Si_9O_{26}(OH)_2$ - *Translucent white to blue, tan; Hardness 5.5; Locality: New Hampshire, USA*

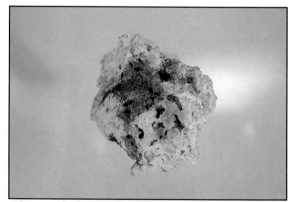

Photography by Jim Hughes, Assisted by ♪ Melody ♫
Collection of Colorado School Of Mines, CO, USA

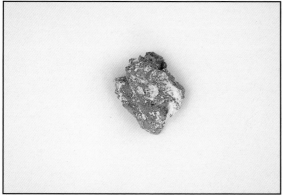

BEAVERITE - $CuPbFe_2(SO_4)_2(OH)_6$ **-** *Earthy yellow; Hardness ?; Locality: Utah, USA*

Photography by Jim Hughes, Assisted by ♪ Melody ♫
Collection of ♪ Melody ♫, Applewood, CO, USA

BENITOITE - $BaTiSi_3O_9$ **-** *Translucent blue, colourless; Hardness 6.3; Locality: California, USA*

Photography by Jim Hughes, Assisted by ♪ Melody ♫
Collection of ♪ Melody ♫, Applewood, CO, USA

BERLINITE - $AlPO_4$ **-** <u>Vitreous</u> *colourless,* <u>greyish</u>*, pale rose; Hardness 6.5; Locality: Skane, Sweden*

Photography by Jim Hughes, Assisted by ♪ Melody ♫
Collection of ♪ Melody ♫, Applewood, CO, USA

BERTHIERITE - $FeSb_2S_4$ -
*Metallic grey; Hardness 2-3;
Locality: Czechoslovakia*

*Photography by Jim Hughes, Assisted by ♪ Melody ♫
Collection of ♪ Melody ♫, Applewood, CO, USA*

BERTRANDITE - *(Crystals,
with blue/purple Fluorapatite)* -
$Be_4Si_2O_7(OH)_2$ - *Vitreous pearly
colourless, pale yellow;
Hardness 6-7; Locality: Minas
Gerais, Brasil*

*Photography by Jim Hughes, Assisted by ♪ Melody ♫
Collection of ♪ Melody ♫, Applewood, CO, USA*

BERYL - $Be_3Al_2Si_6O_{18}$ - *Vitreous
gold; Hardness 7.5-8; Locality:
Minas Gerais, Brasil*

*Photography by Jim Hughes, Assisted by ♪ Melody ♫
Collection of ♪ Melody ♫, Applewood, CO, USA*

BERYLLONITE - $NaBePO_4$ - *Vitreous white, pale yellowish; Hardness 5.5-6; Locality: Maine, USA*

Photography by Jim Hughes, Assisted by ♪ Melody ♫
Collection of Colorado School Of Mines, CO, USA

BERZELIITE - *(Yellow, in black Hausmannite)*- $NaCa_2(Mg,Mn)_2(AsO_4)_3$ - *Resinous yellowish; Hardness 4.5-5; Locality: Sweden*

Photography by Jim Hughes, Assisted by ♪ Melody ♫
Collection of ♪ Melody ♫, Applewood, CO, USA

BETA QUARTZ - *A variety of Quartz formed at temperatures ranging from 573-870 degrees as hexagonals and trapezohedral-hemihedrals; Shows regular hexagonal bi-pyramids with/out a subordinate prism face; Hardness 7; Locality: Mexico*

Photography by Jim Hughes, Assisted by ♪ Melody ♫
Collection of ♪ Melody ♫, Applewood, CO, USA

BETAFITE - *A Niobate and Tintanate of Uranium, etc.; Colour green-black opaque lustrous; Hardness 5: Locality: Ontario, Canada*

Photography by Jim Hughes, Assisted by ♪ Melody ♫
Collection of ♪ Melody ♫, Applewood, CO, USA

BEUDANTITE - $PbFe_3(AsO_4, SO_4)_2(OH)_6$ - *Vitreous resinous black, dark green, brown; Hardness 3.5-4.5; Locality: Utah, USA*

Photography by Jim Hughes, Assisted by ♪ Melody ♫
Collection of ♪ Melody ♫, Applewood, CO, USA

BIEBERITE - $CoSO_4 \cdot 7H_2O$ - *Vitreous red; Hardness 2; Locality: Hessen, Germany*

Photography by Jim Hughes, Assisted by ♪ Melody ♫
Collection of Colorado School Of Mines, CO, USA

BIOTITE - LENS -
$K(Mg,Fe)_3(Si_3Al)O_{10}(OH,F)_2$ -
Black, dark brown, reddish-brown; Hardness 2.5-3; Locality: Portugal

Photography by Jim Hughes, Assisted by ♪ Melody ♫
Collection of ♪ Melody ♫, Applewood, CO, USA

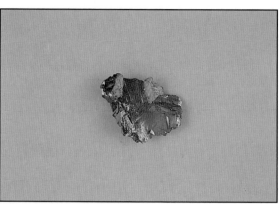

BISMUTH - NATIVE - Bi -
Metallic reddish-white, creamy white in reflected light; Hardness 2-2.5; Locality: Ural Mts., Russia

Photography by Jim Hughes, Assisted by ♪ Melody ♫
Collection of ♪ Melody ♫, Applewood, CO, USA

BITYITE -
$LiCaAl_2(Si_2BeAl)O_{10}(OH)_2$ -
Transparent colourless; Hardness 5.5; Locality: New Mexico, USA

Photography by Jim Hughes, Assisted by ♪ Melody ♫
Collection of ♪ Melody ♫, Applewood, CO, USA

BIXBITE - $Be_3Al_2Si_6O_{18}$ -
Transparent to translucent with
vitreous luster and colour red;
Hardness 7.5-8; Locality:
Colorado, USA

Photography by Jim Hughes, Assisted by ♪ Melody ♫
Collection of Bob Jackson, Applewood, CO, USA

BIXBYITE - Mn_2O_3 -
Metallic/sub-metallic black;
Hardness 6-6.5; Locality: Utah,
USA

Photography by Jim Hughes, Assisted by ♪ Melody ♫
Collection of Bob Jackson, Applewood, CO, USA

BLOEDITE -
$Na_2Mg(SO_4)_2$♥$4H_2O$ - Vitreous
colourless, bluish-green,
reddish; Hardness 2.5-3;
Locality: California, USA

Photography by Jim Hughes, Assisted by ♪ Melody ♫
Collection of Colorado School Of Mines, CO, USA

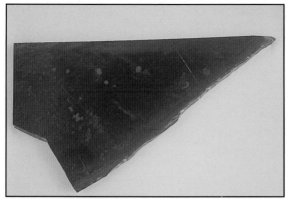

BLOODSTONE - *SiO₂ with impurities* - *Slightly translucent to sub-translucent lustrous green with small spots of red jasper; Hardness 7; Locality: Columbia River, Oregon, USA*

Photography by Jim Hughes, Assisted by ♪ Melody ♫
Collection of Bob Jackson, Applewood, CO, USA
Gift of Howard Dolph, Rufus, Oregon, USA

BOJI STONE - *A brown to brownish-black concretion containing pyrite, palladium and other undefined minerals; Hardness ?; Locality: Western USA*

Photography by Jim Hughes, Assisted by ♪ Melody ♫
Collection of ♪ Melody ♫, Applewood, CO, USA

BOLEITE - $Ag_9Cu_{24}Pb_{26}Cl_{62}(OH)_{48}$ - *Vitreous blue; Hardness 3-3.5; Locality: Mexico*

Photography by Jim Hughes, Assisted by ♪ Melody ♫
Collection of ♪ Melody ♫, Applewood, CO, USA

BOLIVINAIITE - (With Dolomite) Chemical composition not available to author; Hardness 3.5; Colour ranges from white-blue to black; Resinous/vitreous; Locality: Switzerland

Photography by Jim Hughes, Assisted by ♪ Melody ♫
Collection of Colorado School Of Mines, CO, USA

BOOTHITE - $CuSO_4♥7H_2O$ - Silky pearly blue; Hardness from 2-2.5; Locality: Chile, South America

Photography by Jim Hughes, Assisted by ♪ Melody ♫
Collection of ♪ Melody ♫, Applewood, CO, USA

BORNITE - Cu_5FeS_4 - Metallic copper-red/brown, pinkish-brown; Hardness 3; Locality: Mexico

Photography by Jim Hughes, Assisted by ♪ Melody ♫
Collection of ♪ Melody ♫, Applewood, CO, USA

BORNITE - *(With Pyrite)* -
Cu_5FeS_4 - *Metallic
copper-red/brown, pinkish-brown;
Hardness 3; Locality: Mexico*

*Photography by Jim Hughes, Assisted by ♪ Melody ♫
Collection of ♪ Melody ♫, Applewood, CO, USA*

BOTRYOGEN -
$MgFe(SO_4)_2(OH)$♥$7H_2O$ - *Vitreous
light/dark orange-red; Hardness
2-2.5; Locality: Chile, South
America*

*Photography by Jim Hughes, Assisted by ♪ Melody ♫
Collection of ♪ Melody ♫, Applewood, CO, USA*

BOULANGERITE - $5PbS$♥$2Sb_2S_3$
- *Metallic lustrous bluish-grey
crystals and massive; Hardness
2.5-3; Locality: Peru*

*Photography by Jim Hughes, Assisted by ♪ Melody ♫
Collection of ♪ Melody ♫, Applewood, CO, USA*

BRANDTITE -
$Ca_2(Mn,Mg)(AsO_4)_2 \cdot 2H_2O$ -
(In Black Franklinite); Vitreous
colourless, <u>white</u>; Hardness 3.5;
Locality: New Jersey

Photography by Jim Hughes, Assisted by ♪ Melody ♫
Collection of ♪ Melody ♫, Applewood, CO, USA

BRAUNITE - Mn_7SiO_{12} -
(Shiny) Sub-metallic black, grey;
Hardness 6-6.5; Locality:
Germany

Photography by Jim Hughes, Assisted by ♪ Melody ♫
Collection of Colorado School Of Mines, CO, USA

BRAVOITE - $(Fe,Ni)S_2$ - *Pale*
yellow, reddish tarnish;
Hardness 6-6.5; Locality:
Germany

Photography by Jim Hughes, Assisted by ♪ Melody ♫
Collection of Colorado School Of Mines, CO, USA

BRAZILIANITE - *NaAl₃(PO₄)₂(OH)₄ - Vitreous yellow; Hardness 5.5; Locality: Minas Gerais, Brasil*

Photography by Jim Hughes, Assisted by ♪ Melody ♫
Collection of ♪ Melody ♫, Applewood, CO, USA

BRONZITE - *MgSiO₃ - Sub-metallic to adamantine bronze in colour; Hardness 5.5; Locality: Bavaria*

Photography by Jim Hughes, Assisted by ♪ Melody ♫
Collection of ♪ Melody ♫, Applewood, CO, USA

BROOKITE - *TiO₂ - Metallic adamantine brown, black; Hardness 5.5-6; Locality: Arkansas, USA*

Photography by Jim Hughes, Assisted by ♪ Melody ♫
Collection of ♪ Melody ♫, Applewood, CO, USA

BUNSENITE - NiO - Vitreous green; Hardness 5.5; Locality: Saxony, Germany

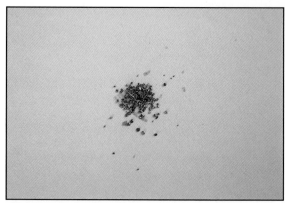

Photography by Jim Hughes, Assisted by ♪ Melody ♫
Collection of ♪ Melody ♫, Applewood, CO, USA

BUSTAMITE - CaMnSi$_2$O$_6$ - Translucent pale pink, brownish-red; Hardness 5.5-6.5; Republic of South Africa.

Photography by Jim Hughes, Assisted by ♪ Melody ♫
Collection of Bob Jackson, Applewood, CO, USA

BUTLERITE - FeSo$_4$(OH)♥2H$_2$O - Vitreous orange to yellow-orange; Hardness 2.5; Locality: Chile, South America

Photography by Jim Hughes, Assisted by ♪ Melody ♫
Collection of ♪ Melody ♫, Applewood, CO, USA

CAFARSITE -
$(Ca,Mn)_8(Ti,Fe)_{6.5}(AsO_3)_{12} \heartsuit 2H_2O$
- *Translucent dark brown;*
Hardness 5.5-6; Locality:
Valais, Switzerland.

Photography by Jim Hughes, Assisted by ♪ Melody ♫
Collection of Colorado School Of Mines, CO, USA

CALAVERITE - $AuTe_2$ -
(Silvery) Metallic yellow/white,
creamy-white in reflected light;
Hardness 2.5-3; Locality:
Colorado, USA

Photography by Jim Hughes, Assisted by ♪ Melody ♫
Collection of ♪ Melody ♫, Applewood, CO, USA

CALCITE - *(Over Quartz)* -
$CaCO_3$ - *Vitreous colourless,*
white, yellow, pink, purple,
green, peach, blue, etc.;
Hardness 3; Locality: Republic
of South Africa, Africa

Photography by Jim Hughes, Assisted by ♪ Melody ♫
Collection of ♪ Melody ♫, Applewood, CO, USA

CALCITE - *(Dog-toothed Elestial); $CaCO_3$ - Vitreous colourless, white, yellow, pink, purple, green, peach, blue, etc.; Hardness 3; Locality: Mexico*

Photography by Jim Hughes, Assisted by ♪ Melody ♫
Collection of ♪ Melody ♫, Applewood, CO, USA

CALCITE - *(Pink-Peach); $CaCO_3$; Vitreous; Hardness 3; Locality: Mexico*

Photography by Jim Hughes, Assisted by ♪ Melody ♫
Collection of ♪ Melody ♫, Applewood, CO, USA

CALCITE - *(Pink, Peach, and Iceland Spar); $CaCO_3$; Vitreous; Hardness 3; Locality: Mexico and Iceland*

Photography by Jim Hughes, Assisted by ♪ Melody ♫
Collection of ♪ Melody ♫, Applewood, CO, USA

CALCITE - (Gold); $CaCO_3$;
Vitreous; Hardness 3; Locality:
Mexico

Photography by Jim Hughes, Assisted by ♪ Melody ♫
Collection of ♪ Melody ♫, Applewood, CO, USA

CALCITE - (With Copper
Inclusions); $CaCO_3$; Vitreous;
Hardness 3; Locality: Namibia,
Africa

Photography by Jim Hughes, Assisted by ♪ Melody ♫
Collection of ♪ Melody ♫, Applewood, CO, USA

CALCITE - (Gold - Green);
$CaCO_3$; Vitreous; Hardness 3;
Locality: Mexico

Photography by Jim Hughes, Assisted by ♪ Melody ♫
Collection of ♪ Melody ♫, Applewood, CO, USA
Gifts of W.R. Horning and Margarite Martin

CALEDONITE *(With Linarite) -*
$Cu_2Pb_5(SO_4)_3(CO_3)(OH)_6$ *-*
Resinous green, bluish-green;
Hardness 2.5-3; Locality:
Lanarkshire, Scotland.1067-3

Photography by Jim Hughes, Assisted by ♪ Melody ♫
Collection of ♪ Melody ♫, Applewood, CO, USA

CALOMEL *- alpha-HgCl;*
Adamantine colourless, white,
grey, yellow, brown to pinkish-
brown; Hardness 1.5; Texas, USA

Photography by Jim Hughes, Assisted by ♪ Melody ♫
Collection of Colorado School Of Mines, CO, USA

CANCRINITE *-*
$(Na,Ca)_8(Si_6Al_6)O_{24}(CO_3)_2$*; Sub-*
vitreous pearly white to pink, grey,
yellow, green blue, etc.; Hardness
5-6; Locality: Ontario, Canada

Photography by Jim Hughes, Assisted by ♪ Melody ♫
Collection of ♪ Melody ♫, Applewood, CO, USA

CAPPELENITE -
$Ba(Y,Ce)_6B_6Si_3O_{24}F_2$ -
Vitreous/lustrous <u>*greenish-*</u>
<u>*brown;*</u> *Hardness 6-6.5;*
Locality: Langesundfjord,
Norway

CARNALLITE - *KMgCl₃♥6H₂O;*
Greasy colourless, white, <u>*orange*</u>
<u>*to reddish;*</u> *Hardness 2.5;*
Saxony, Germany.

CARNELIAN - SiO_2 *with*
impurities; Transparent to
translucent lustrous pale to deep
red, brownish-red, red-to-brown;
Hardness 7; Locality: Messina
area, Republic of South Africa,
Africa

CARNELIAN - SiO₂ with impurities; Transparent to translucent lustrous pale to deep red, brownish-red, red-to-brown; Hardness 7; Locality: Oregon, USA
Photography by Jim Hughes, Assisted by ♪ Melody ♫
Collection of Bob Jackson, Applewood, CO, USA

CARNOTITE - (Yellow) - K₂O♥2UO₃♥V₂O₅♥2H₂O; Hardness 2-2.5; Locality:
   Western Australia
Photography by Jim Hughes, Assisted by ♪ Melody ♫
Collection of ♪ Melody ♫, Applewood, CO, USA

CARROLLITE - $Cu(Co,Ni)_2S_4$ - (Silvery) Metallic grey, white in reflected light; Hardness 4.5-5.5; Locality: Maryland, USA

Photography by Jim Hughes, Assisted by ♪ Melody ♫
Collection of ♪ Melody ♫, Applewood, CO, USA

CASSITERITE - SnO_2 - Adamantine metallic brown, yellow, grey, white, etc.,; Hardness 6-7; Locality: Minas, Gerais, Brasil

Photography by Jim Hughes, Assisted by ♪ Melody ♫
Collection of ♪ Melody ♫, Applewood, CO, USA

CATLINITE - Al_2SiO_5♥nH_2O with iron impurities; Vitreous to sub/resinous luster with red-to-brown, brownish-pink colour; Hardness 3; Minnesota, USA

Photography by Jim Hughes, Assisted by ♪ Melody ♫
Collection of ♪ Melody ♫, Applewood, CO, USA

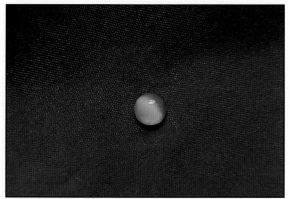

CAT'S EYE - *(Quartz variety, which is SiO_2 with possible fibers of asbestos); Vitreous luster with greenish/goldish, etc., chatoyant colours; Hardness 8.5; Locality, Minas Gerais, Brasil*

CELESTITE - $SrSO_4$ - *Vitreous to pearly <u>blue</u>, orange ranges; Hardness 3-3.5; Locality: Madagascar*

CELESTITE - $SrSO_4$ - *Vitreous to pearly blue, <u>orange</u> ranges; Hardness 3-3.5; Locality: Canada*

CERULEITE -
$Cu_2Al_7(AsO_4)_4(OH)_{13}$♥$12H_2O$ -
Clayish turquoise-blue;
Hardness 5-6; Locality:
Huanaco, Chile.

Photography by Jim Hughes, Assisted by ♪ Melody ♫
Collection of ♪ Melody ♫, Applewood, CO, USA

CERULEITE - (Slice)
$Cu_2Al_7(AsO_4)_4(OH)_{13}$♥$12H_2O$ -
Clayish turquoise-blue;
Hardness 5-6; Locality:
Huanaco, Chile.

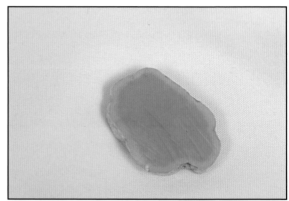

Photography by Jim Hughes, Assisted by ♪ Melody ♫
Collection of ♪ Melody ♫, Applewood, CO, USA

CERUSSITE - $PbCO_3$ -
Adamantine vitreous yellow,
colourless, white, grey;
Hardness 3-3.5; Locality:
Namibia, Africa

Photography by Jim Hughes, Assisted by ♪ Melody ♫
Collection of Bob Jackson, Applewood, CO, USA

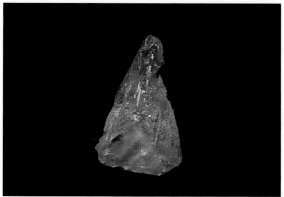

CERUSSITE - PbCO₃ -
Adamantine vitreous <u>yellow</u>,
colourless, white, grey; Hardness
3-3.5; Locality: Namibia, Africa

Photography by Jim Hughes, Assisted by ♪ Melody ♫
Collection of ♪ Melody ♫, Applewood, CO, USA

CHABAZITE -
$Ca(Al_2Si_4)O_{12}$♥$6H_2O$ - *Vitreous*
<u>white</u>, pink, brick-red; Hardness
4-5; Locality: Oregon, USA

Photography by Jim Hughes, Assisted by ♪ Melody ♫
Collection of Bob Jackson, Applewood, CO, USA

CHALCANTHITE - *Vitreous*
Berlin-blue to sky-blue;
$CuSO_4$♥$5H_2O$; *Hardness 2.5;*
Locality: Arizona, USA

Photography by Jim Hughes, Assisted by ♪ Melody ♫
Collection of Bob Jackson, Applewood, CO, USA

CHALCANTHITE - *Vitreous Berlin-blue to sky-blue;* $CuSO_4 \cdot 5H_2O$; *Hardness 2.5; Locality: Arizona*

Photography by Jim Hughes, Assisted by ♪ Melody ♫
Collection of John Wittman, Wheat Ridge, CO, USA

CHALCEDONY - *(After Aragonite) -* SiO_2 - *Lustrous transparent to translucent white, <u>red</u>, greyish, blue, brown, black, gold/yellow, etc.; Hardness 7; Locality: Texas, USA*

Photography by Jim Hughes, Assisted by ♪ Melody ♫
Collection of ♪ Melody ♫, Applewood, CO, USA

CHALCOPYRITE - *With Pyrite;* $20CuO \cdot Al_2O_3 \cdot 2As_2O_5 \cdot 3SO_3 \cdot 25H_2O$; *Vitreous/sub-adamantine greens, etc.; Locality: Arizona, USA*

Photography by Jim Hughes, Assisted by ♪ Melody ♫
Collection of ♪ Melody ♫, Applewood, CO, USA

CHALCOSIDERITE -
$CuFe_6(PO_4)_4(OH)_8$♥$4H_2O$ -
*Vitreous light green hydrous
copper-iron phosphate; Hardness
4.5; Locality: Cornwall, England*

*Photography by Jim Hughes, Assisted by ♪ Melody ♫
Collection of Colorado School Of Mines, CO, USA*

CHAROITE -
$(K,Na)_5(Ca,Ba,Sr)_8Si_{18}O_{46}(OH,F)$♥
nH_2O - *Translucent to opaque
violet; Hardness 5; Locality:
Yakutiya, Russia.*

*Photography by Jim Hughes, Assisted by ♪ Melody ♫
Collection of Bob Jackson, Applewood, CO, USA*

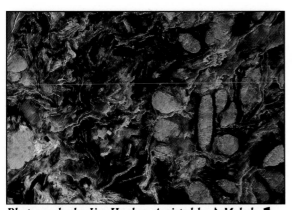

CHAROITE - *(With Quartz
Crystals); Translucent to opaque
violet; Hardness 5;*
$(K,Na)_5(Ca,Ba,Sr)_8Si_{18}O_{46}(OH,F)$♥
nH_2O; *Locality: Yakutiya, Russia.*

*Photography by Jim Hughes, Assisted by ♪ Melody ♫
Collection of ♪ Melody ♫, Applewood, CO, USA*

CHERT - SiO_2 - *Opaque grey, brown, <u>black</u> with sub-vitreous luster with brittleness and impurities; Hardness 7; Locality: Ohio, USA where???*

Photography by Jim Hughes, Assisted by ♪ Melody ♫
Collection of Bob Jackson, Applewood, CO, USA

CHIASTOLITE - $AlSiO_5$ *with carbonaceous impurities - Vitreous luster with tesselation in transverse, whitish, rose-to-flesh red, violet, pearl-grey, reddish-brown, olive-green colours with tesselation contrasting; Hardness 7.5; Locality: Spain*

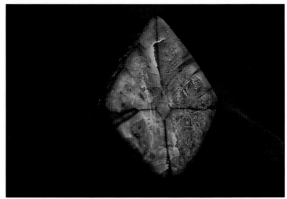

Photography by Jim Hughes, Assisted by ♪ Melody ♫
Collection of ♪ Melody ♫, Applewood, CO, USA

CHILDRENITE - $(Fe,Mn)AlPO_4(OH)_2 ♥ H_2O$ - *Vitreous resinous <u>brown</u>, yellowish-brown; Hardness 5; Locality: Yukon Territory, Canada*

Photography by Jim Hughes, Assisted by ♪ Melody ♫
Collection of ♪ Melody ♫, Applewood, CO, USA

CHINESE WRITING ROCK -
Basalt porphyry with feldspar crystals; Hardness variable; Locality: California

Photography by Jim Hughes, Assisted by ♪ Melody ♫
Collection of Bob Jackson, Applewood, CO, USA

CHLORITE -
$(Mg,Al,Fe,Li,Mn,Ni)_{4-6}$ $(Si,Al,B,Fe)_4O_{10}(OH,O)_8$ *[General sheet silicate] - (Serefina) green in photograph; Hardness 1-4; Locality: Germany*

Photography by Jim Hughes, Assisted by ♪ Melody ♫
Collection of ♪ Melody ♫, Applewood, CO, USA
Gift from Ms Sigari, Germany

CHLOROCALCITE - $KCaCl_3$ **-**
Transparent to translucent white (crust); Hardness 2.5-3; Locality: Campania, Italy

Photography by Jim Hughes, Assisted by ♪ Melody ♫
Collection of ♪ Melody ♫, Applewood, CO, USA

"CHOCOLATE MARBLES" -
Petrified volcanic mud balls;
Hardness ?; Arizona, USA.

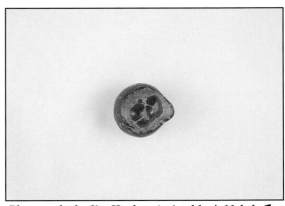

Photography by Jim Hughes, Assisted by ♪ Melody ♫
Collection of ♪ Melody ♫, Applewood, CO, USA

CHROMITE - *FeCr$_2$O$_4$* -
Translucent to <u>opaque iron-</u>
<u>black</u>, brownish-black,
yellowish-red, greenish, etc.;
Hardness 5.5; Locality: Ukraine

Photography by Jim Hughes, Assisted by ♪ Melody ♫
Collection of ♪ Melody ♫, Applewood, CO, USA

CHRYSOBERYL - *BeAl$_2$O$_4$* -
Vitreous green, yellow, <u>yellow to</u>
<u>green-yellow</u>, greenish-brown;
Hardness 8.5; Locality: Minas
Gerais, Brasil

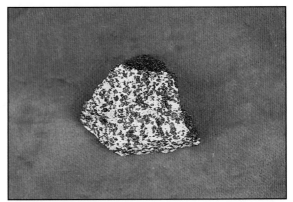

Photography by Jim Hughes, Assisted by ♪ Melody ♫
Collection of ♪ Melody ♫, Applewood, CO, USA

CHRYSOCOLLA - *(Gemmy)* - $(Cu,Al)_2H_2Si_2O_5(OH)_4♥nH_2O$ - *Vitreous earthy green,*
  bluish-green, blue; Hardness 2-4; Locality: Arizona, USA
Photography by Jim Hughes, Assisted by ♪ Melody ♫
Collection of ♪ Melody ♫, Applewood, CO, USA
Gift from Bob Jackson, Applewood, CO, USA

CHRYSOCOLLA - *(Drusy)* - $(Cu,Al)_2H_2Si_2O_5(OH)_4♥nH_2O$ - *Vitreous earthy green,*
  bluish-green, blue; Hardness 2-4; Locality: Arizona, USA
Photography by Jim Hughes, Assisted by ♪ Melody ♫
Collection of ♪ Melody ♫, Applewood, CO, USA

CHRYSOPRASE - SiO₂ with nickel oxide impurities; Lustrous transparent to translucent apple-green; Hardness 7: Locality: Australia

Photography by Jim Hughes, Assisted by ♪ Melody ♫
Collection of Bob Jackson, Applewood, CO, USA

CHRYSOTILE - Sub-resinous lustrous green-black, brown-red, brown-yellow, white; hydrated silicate; Hardness 2.5-4; Locality: Arizona, USA

Photography by Jim Hughes, Assisted by ♪ Melody ♫
Collection of ♪ Melody ♫, Applewood, CO, USA

CINNABAR - (With Stibnite - [black]); HgS; Translucent to opaque adamantine red; Hardness 5-6; Locality: Alaska, USA

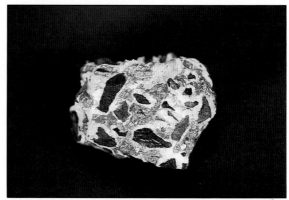

Photography by Jim Hughes, Assisted by ♪ Melody ♫
Collection of ♪ Melody ♫, Applewood, CO, USA

CINNABAR - (With Steatite - [cream]); HgS; Translucent to opaque adamantine to deep red; Hardness 5-6; Locality: Western USA

Photography by Jim Hughes, Assisted by ♪ Melody ♫
Collection of ♪ Melody ♫, Applewood, CO, USA

CITRINE - SiO₂ with colloidal iron hydrates - Yellow to gold to brownish-gold quartz; Hardness 7; Locality: Minas Gerais, Brasil

Photography by Jim Hughes, Assisted by ♪ Melody ♫
Collection of ♪ Melody ♫, Applewood, CO, USA

CITRINE - (With Hematite) - SiO₂ with colloidal iron hydrates; Yellow to gold to brownish-gold quartz; Hardness 7; Locality: Minas Gerais, Brasil

Photography by Jim Hughes, Assisted by ♪ Melody ♫
Collection of Bob Jackson, Applewood, CO, USA

CLEVELANDITE - *(With Fluorite and Quartz);* $NaAlSi_3O_8$; *Lamellar masses, with calcium usually present as* $CaAl_2Si_2O_8$; *Hardness 6-6.5; Locality:Pakistan*

Photography by Jim Hughes, Assisted by ♪ Melody ♫
Collection of ♪ Melody ♫, Applewood, CO, USA

CLINOCHLORE - $(Mg,Al)_6(SiAl)_4O_{10}(OH)_8$ - *Pearly* <u>green</u>, *olive-green, yellowish, white; Hardness 2-2.5; Locality: Canton Wallis, Switzerland*

Photography by Jim Hughes, Assisted by ♪ Melody ♫
Collection of ♪ Melody ♫, Applewood, CO, USA

CLINOZOISITE - *(With Quartz)* $HCa_2Al_3Si_3O_{13}$ - *Contains less than 10 percent of the iron molecule and is optically positive, distinguishing it from Zoisite; Hardness 6-6.5; Locality: Nevada, USA*

Photography by Jim Hughes, Assisted by ♪ Melody ♫
Collection of ♪ Melody ♫, Applewood, CO, USA

COBALTITE - *CoAsS; Silver-white inclined to red, steel-grey with violet tinge, greyish-black; Hardness 5.5; Locality: Zaire, Africa*

COLEMANITE - $CaB_3O_4(OH)_3 \cdot H_2O$ - *Vitreous adamantine colourless, white, yellowish, grey; Hardness 4.5; Locality: California, USA*

COLUMBITE - *Niobate and tantalate of iron and manganese $[(Fe,Mn)(Nb,Ta)_2O_6]$; Lustrous, sub-metallic, sub-resinous opaque iron-black, grey and brown-black, reddish-brown (translucent), sometimes iridescent; Hardness 6; Locality: Minas Gerais, Brasil*

CONCRETION - Various compositions - Creation via deposition of mineral(s) via aqueous solution within dissimilar mineral with coalescence of separate particles of matter into one formation; Hardness variable; Locality: Northern Mid-Western USA

Photography by Jim Hughes, Assisted by ♪ Melody ♫
Collection of ♪ Melody ♫, Applewood, CO, USA
Gift of Ed Maslovicz, Sanctuary Crystals, Alsip, IL, USA
(A "Old" Stone For New Objectives)

CONCRETION - Various compositions - Creation via deposition of mineral(s) via aqueous solution within dissimilar mineral with coalescence of separate particles of matter into one formation; Hardness variable; Locality: Northern Mid-Western USA

Photography by Jim Hughes, Assisted by ♪ Melody ♫
Collection of ♪ Melody ♫, Applewood, CO, USA

CONICHALCITE - $CaCuAsO_4(OH)$ - Vitreous green, yellowish-green; Hardness 4.5; Utah, USA

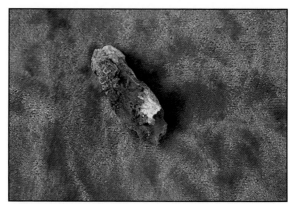

Photography by Jim Hughes, Assisted by ♪ Melody ♫
Collection of ♪ Melody ♫, Applewood, CO, USA

COOKEITE - *(On Quartz)* -
$(Al,Li)_3Al_2(Si,Al)_4O_{10}(OH)_8$ -
*Pearly white, <u>green</u>, yellow-gold,
yellowish-green; Hardness 2.5;
Locality: Arkansas, USA*

Photography by Jim Hughes, Assisted by ♪ Melody ♫
Collection of ♪ Melody ♫, Applewood, CO, USA

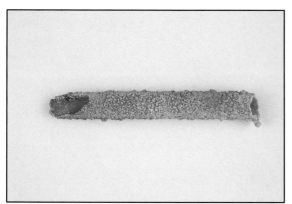

COOKEITE - *(Cast)* -
$(Al,Li)_3Al_2(Si,Al)_4O_{10}(OH)_8$ -
*Pearly white, green, <u>yellow-gold</u>,
yellowish-green; Hardness 2.5;
Locality: Minas Gerais, Brasil*

Photography by Jim Hughes, Assisted by ♪ Melody ♫
Collection of ♪ Melody ♫, Applewood, CO, USA

COPAL - *An oxygenated
hydrocarbon, resinous, ranging in
colour from yellow to red to
brown; Hardness 2-2.5; Locality:
Latin America*

Photography by Jim Hughes, Assisted by ♪ Melody ♫
Collection of Bob Jackson, Applewood, CO, USA

COPPER - (Native); Cu; Metallic light rose, copper-red; Hardness 2.5-3; Locality: Michigan, USA

Photography by Jim Hughes, Assisted by ♪ Melody ♫
Collection of ♪ Melody ♫, Applewood, CO, USA

COPPER - (With Quartz); Cu; Metallic light rose, copper-red; Hardness 2.5-3; Locality: Messina, Republic of South Africa, Africa

Photography by Jim Hughes, Assisted by ♪ Melody ♫
Collection of ♪ Melody ♫, Applewood, CO, USA

CORAL - (White) - Fossilized calcium carbonate with impurities; Hardness variable; Locality: Brasil

Photography by Jim Hughes, Assisted by ♪ Melody ♫
Collection of ♪ Melody ♫, Applewood, CO, USA
Gift of Pai Carumbe', Belo Horizonte, Minas Gerais, Brasil

CORAL - (Red) - Calcium carbonate with impurities; Hardness variable; Locality: Mediterranean Sea

Photography by Jim Hughes, Assisted by ♪ Melody ♫
Collection of ♪ Melody ♫, Applewood, CO, USA

CORAL - (Black) - Calcium carbonate with impurities; Hardness variable; Locality: Mexican Gulf Coast

Photography by Jim Hughes, Assisted by ♪ Melody ♫
Collection of ♪ Melody ♫, Applewood, CO, USA

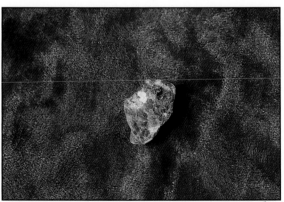

CORDIERITE - $Mg_2Al_4Si_5O_{18}$ - Translucent greyish, lilac, blue; Hardness 7; Locality: India

Photography by Jim Hughes, Assisted by ♪ Melody ♫
Collection of ♪ Melody ♫, Applewood, CO, USA

CORNETITE - $Cu_3PO_4(OH)_3$ - *Vitreous blue, greenish-blue; Hardness 4.5; Locality: Shaba, Zaire*

Photography by Jim Hughes, Assisted by ♪ Melody ♫
Collection of ♪ Melody ♫, Applewood, CO, USA

CORUNDUM - Al_2O_3 - *Adamantine vitreous colourless, blue, red, brown-red, etc.; Hardness 9; Locality: Sri Lanka*

Photography by Jim Hughes, Assisted by ♪ Melody ♫
Collection of Bob Jackson, Applewood, CO, USA

CORUNDUM - *(With Mica [grey-blue])* Al_2O_3 - *Vitreous adamantine colourless, blue, red, brown-red, etc.; Hardness of 9; Locality: Montana, USA*

Photography by Jim Hughes, Assisted by ♪ Melody ♫
Collection of Bob Jackson, Applewood, CO, USA

COVELLITE - *CuS; Sub-metallic blue; Hardness 1.5-2; Locality: Montana, USA*

Photography by Jim Hughes, Assisted by ♪ Melody ♫
Collection of Bob Jackson, Applewood, CO, USA

COVELLITE - *(With Pyrite) - CuS; Sub-metallic blue; Hardness 1.5-2; Locality: Montana, USA*

Photography by Jim Hughes, Assisted by ♪ Melody ♫
Collection of ♪ Melody ♫, Applewood, CO, USA

CRANDALLITE - *$CaO♥2Al_2O_3♥P_2O_5♥5H_2O$ with phenocrysts ("flowers" in the stone) which have caused the formation to be called "Chrysanthemum Stone"; Hardness 4 (Crandallite); Locality: Belgium*

Photography by Jim Hughes, Assisted by ♪ Melody ♫
Collection of ♪ Melody ♫, Applewood, CO, USA

CREEDITE -
$Ca_3Al_2SO_4(OH)_2F_8 \heartsuit 2H_2O$ -
Vitreous colourless, white purple; Hardness 4; Locality: Mexico

Photography by Jim Hughes, Assisted by ♪ Melody ♫
Collection of ♪ Melody ♫, Applewood, CO, USA

CRISTOBALITE - SiO_2 -
Vitreous white; Hardness 7; Locality: Sonora, Mexico

Photography by Jim Hughes, Assisted by ♪ Melody ♫
Collection of ♪ Melody ♫, Applewood, CO, USA

CROCOITE - $PbCrO_4$ -
Adamantine vitreous red, orange, yellow; Hardness 2.5-3; Locality: Ural Mts., Russia.

Photography by Jim Hughes, Assisted by ♪ Melody ♫
Collection of ♪ Melody ♫, Applewood, CO, USA

CRYOLITE - *alpha-Na₃AlF₆* -
*Vitreous lustrous white, brownish,
greyish, black; Hardness 2.5;
Ivigtut, Greenland*

*CRYOLITE - alpha-Na_3AlF_6 -
Vitreous lustrous white, brownish,
greyish, black; Hardness 2.5;
Ivigtut,   Greenland*

Photography by Jim Hughes, Assisted by ♪ Melody ♫
Collection of ♪ Melody ♫, Applewood, CO, USA

*CUBANITE - $CuFe_2S_3$ - Metallic
brass, bronze-yellow; Hardness
3.5; Locality: Barracanao, Cuba*

Photography by Jim Hughes, Assisted by ♪ Melody ♫
Collection of ♪ Melody ♫, Applewood, CO, USA

*CUMBERLANDITE -
Composition undefined in
conventional sources; Hardness
tested between 4-6; Locality:
Eastern USA*

Photography by Jim Hughes, Assisted by ♪ Melody ♫
Collection of ♪ Melody ♫, Applewood, CO, USA

CUPRITE - Cu_2O - *Adamantine sub-metallic red; Hardness from 3.5-4; Locality: Arizona, USA*

Photography by Jim Hughes, Assisted by ♪ Melody ♫
Collection of ♪ Melody ♫, Applewood, CO, USA

CUPROADAMITE (Green) - $Zn_3As_2O_8$♥$Zn(OH)_2$ with Copper; *Hardness 4; Locality: Namibia, Africa*

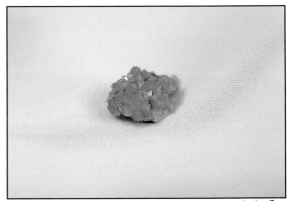

Photography by Jim Hughes, Assisted by ♪ Melody ♫
Collection of ♪ Melody ♫, Applewood, CO, USA

CUSPIDINE - $Ca_4Si_2O_7(F,OH)_2$ - *Vitreous green spear-shaped crystals; Hardness 5-6; Locality: Campania, Italy*

Photography by Jim Hughes, Assisted by ♪ Melody ♫
Collection of ♪ Melody ♫, Applewood, CO, USA

CYANOTRICHITE - $4CuO \cdot Al_2O_3 \cdot SO_3 \cdot 8H_2O$; *Velvety druses, blue to colourless; Hardness 2-3; Locality: Arizona, USA*

Photography by Jim Hughes, Assisted by ♪ Melody ♫
Collection of ♪ Melody ♫, Applewood, CO, USA

CYLINDRITE - *(Crystals in Franckeite and Teallite) -* $FePb_3Sn_4Sb_2S_{14}$; *Metallic blackish-grey, grey-white in reflected light; Hardness 2.5; Locality: Bolivia, South America*

Photography by Jim Hughes, Assisted by ♪ Melody ♫
Collection of ♪ Melody ♫, Applewood, CO, USA

CYMOPHANE - $BeAl_2O_4$; *Vitreous luster with greenish/goldish, brownish, reddish chatoyant/opalescent colours; Hardness 8.5; Locality: Minas Gerais, Brasil*

Photography by Jim Hughes, Assisted by ♪ Melody ♫
Collection of ♪ Melody ♫, Applewood, CO, USA

Our star shall light the path for us
And we will journey far, in trust

[Julianne Guilbault]

DAMSONITE - SiO_2 with impurities; Opaque vitreous to sub-resinous, sometimes occurring with crystalline quartz, blue to purple-violet-brown in colour; Hardness 7; Locality: Arizona, USA

Photography by Jim Hughes, Assisted by ♪ Melody ♫
Collection of Bob Jackson, Applewood, CO, USA

DANALITE - $Be_3Fe_4(SiO_4)_3S$ - Vitreous resinous <u>pink</u>, grey; Hardness 5.5-6; Locality: New Hampshire, USA

Photography by Jim Hughes, Assisted by ♪ Melody ♫
Collection of ♪ Melody ♫, Applewood, CO, USA

DANBURITE - $CaB_2Si_2O_8$ - Vitreous lustrous colourless, pale pink, pale yellow, yellowish-brown, pinkish; Hardness 7-7.2; Locality: Mexico

Photography by Jim Hughes, Assisted by ♪ Melody ♫
Collection of ♪ Melody ♫, Applewood, CO, USA

DATOLITE - CaBSiO₄(OH) 0
*Vitreous white, greyish, pale
<u>green</u>, red, yellow, etc.; Hardness
5-5.5; Locality: Mexico*

*Photography by Jim Hughes, Assisted by ♪ Melody ♫
Collection of ♪ Melody ♫, Applewood, CO, USA*

**DAVYNE -$(Na,Ca,K)_8(Si_6Al_6)O_{24}$
$(CL,SO_4,CO_3)_{2-3}$ - Vitreous pearly
colourless, white; Hardness 5.5;
Campania, Italy.**

*Photography by Jim Hughes, Assisted by ♪ Melody ♫
Collection of Colorado School of Mines, CO, USA*

DESCLOISITE -
*$Pb(Zn,Cu)VO_4OH$; <u>Black</u>, red,
brownish-red, brown; Hardness
3.5; Locality: Zimbabwe, Africa*

*Photography by Jim Hughes, Assisted by ♪ Melody ♫
Collection of Bob Jackson, Applewood, CO, USA*

DIABANTITE -
$H_4(Mg,Fe)_2Al_2SiO_9$ - Fine scaly to fibrous and earthy form filling seams in igneous rocks; very deep green to green/gold; Hardness 2.5-3; Locality: Germany

Photography by Jim Hughes, Assisted by ♪ Melody ♫
Collection of ♪ Melody ♫, Applewood, CO, USA

DIASPOR - *Al_2O_3♥H_2O - Brilliant luster, whitish, greyish, green-grey, brown, yellow to colourless; Tarnishes to vitreous yellow; Hardness 6.5-7; Locality: Turkey*

Photography by Jim Hughes, Assisted by ♪ Melody ♫
Collection of ♪ Melody ♫, Applewood, CO, USA
Gift from Ms Sigari, Germany

DICINITE- *PbV_2O_6 - (Also known as Dechenite); Silvery, red, brown, yellow with resinous luster; Hardness 4; Locality: Carinthia, Austria*

Photography by Jim Hughes, Assisted by ♪ Melody ♫
Collection of ♪ Melody ♫, Applewood, CO, USA

DINOSAUR BONE - *Fossilized portion of dinosaur; Hardness variable; Locality: Wyoming, USA*

Photography by Jim Hughes, Assisted by ♪ Melody ♫
Collection of ♪ Melody ♫, Applewood, CO, USA

DIOPSIDE - $CaMgSi_2O_6$ - *Vitreous resinous white, yellowish, greyish, pale green; Hardness 5.5-6.5; Locality: Rajasthan, India*

Photography by Jim Hughes, Assisted by ♪ Melody ♫
Collection of ♪ Melody ♫, Applewood, CO, USA

DIOPTASE - $CuSiO_3 ♥ H_2O$ - *Vitreous green; Hardness 5; Locality: Tsumeb, Namibia.*

Photography by Jim Hughes, Assisted by ♪ Melody ♫
Collection of ♪ Melody ♫, Applewood, CO, USA

DIOPTASE - (Inclusions in Calcite); CuSiO₃♥H₂O - Vitreous green;
Hardness 5; Locality: Tsumeb, Namibia.
Photography by Jim Hughes, Assisted by ♪ Melody ♫
Collection of ♪ Melody ♫, Applewood, CO, USA

DIOPTASE - *(On Calcite)* - $CuSiO_3 \cdot H_2O$ - *Vitreous green; Hardness 5; Locality: Tsumeb, Namibia.*

Photography by Jim Hughes, Assisted by ♪ Melody ♫
Collection of ♪ Melody ♫, Applewood, CO, USA

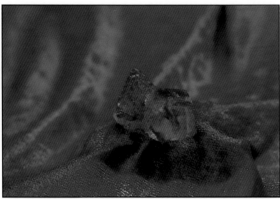

DOLOMITE *(Gemmy)* - $CaMg(CO_3)_2$ - *Vitreous colourless, white, <u>peach</u>, gold, grey, greenish, pinkish, etc.; Hardness 3.5-4; Locality: Mexico*

Photography by Jim Hughes, Assisted by ♪ Melody ♫
Collection of ♪ Melody ♫, Applewood, CO, USA

DOMEYKITE - Cu_3As; *Tin-white to steel-grey, readily tarnished; Hardness 3.5-4; Locality: Michigan, USA*

Photography by Jim Hughes, Assisted by ♪ Melody ♫
Collection of ♪ Melody ♫, Applewood, CO, USA

DOUGLASITE - $K_2FeCl_4 \heartsuit 2H_2O$
- Vitreous light green; Hardness 2.5; Locality: Silesia, Poland

Photography by Jim Hughes, Assisted by ♪ Melody ♫
Collection of ♪ Melody ♫, Applewood, CO, USA

DUFTITE *- $PbCuAsO_4(OH)$ - Vitreous <u>olive-green</u>, grey-green; Hardness 3; Locality: Tsumeb, Namibia*

Photography by Jim Hughes, Assisted by ♪ Melody ♫
Collection of ♪ Melody ♫, Applewood, CO, USA

DUMONTITE *- (With Kasolite); $Pb_2(UO_2)_3(PO_4)_2O_2 \heartsuit 5H_2O$ - Translucent yellow; Hardness from 2-2.5; Locality: Nevada, USA*

Photography by Jim Hughes, Assisted by ♪ Melody ♫
Collection of ♪ Melody ♫, Applewood, CO, USA

DUMORTIERITE - $Al_7O_3(BO_3)(SiO_4)_3$ - *Vitreous blue to violet and pink to brown; Hardness 6.5-8.5; Locality: Arizona, USA*

Photography by Jim Hughes, Assisted by ♪ Melody ♫
Collection of ♪ Melody ♫, Applewood, CO, USA

DUNDASITE - $PbAl_2(CO_3)_2(OH)_4 \heartsuit H_2O$ - *Vitreous silky white; Hardness 2; Locality: Belgium*

Photography by Jim Hughes, Assisted by ♪ Melody ♫
Collection of ♪ Melody ♫, Applewood, CO, USA

DYSCRASITE - Ag_3Sb - *Metallic White; Hardness 3.5-4; Locality: Czechoslovakia*

Photography by Jim Hughes, Assisted by ♪ Melody ♫
Collection of ♪ Melody ♫, Applewood, CO, USA

One Earth, One People,
Together We Are, In The Universe

[Marie Mockers, Kyoto, Japan]

ECKERMANNITE -
$Na_3(Mg,Fe)_4AlSi_8O_{22}(OH)_2$ -
Vitreous dark to bluish-green;
Hardness 5-6; Locality: Granna,
Sweden

Photography by Jim Hughes, Assisted by ♪ Melody ♫
Collection of ♪ Melody ♫, Applewood, CO, USA

EGLESTONITE - Hg_4Cl_2O;
Adamantine to resinous; brown
to yellow, darkening on
exposure to black; Hardness
from 2-3; Locality: California,
USA

Photography by Jim Hughes, Assisted by ♪ Melody ♫
Collection of ♪ Melody ♫, Applewood, CO, USA

EILAT STONE - *Combination*
of Chrysocolla, Turquoise,
Copper, and impurities;
Hardness variable; Locality:
King Solomon's Copper mines
on the Red Sea

Photography by Jim Hughes, Assisted by ♪ Melody ♫
Collection of Bob Jackson, Applewood, CO, USA

EILAT STONE - *Combination of Chrysocolla, Turquoise, Copper, and impurities; Hardness variable; Locality: King Solomon's Copper mines on the Red Sea*

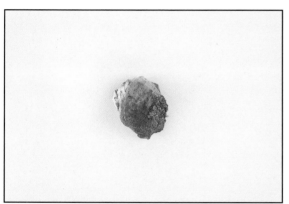

EMBOLITE - *Ag(Br,Cl) with variable ratio of chlorine to bromine - Luster resinous to adamantine with colour ranging from greyish-green to yellowish-green and yellow; Hardness 1-1.5; Locality: Arizona, USA*

EMERALD (Crystal) - $Be_3Al_2Si_6O_{18}$ *with chromium; Transparent to translucent with vitreous luster and colour emerald green; Hardness 7.5-8; Locality: Columbia, South America*

EMMONSITE -
$Fe_2(TeO_3)_3 \cdot 2H_2O$ - *Vitreous yellowish-green; Hardness 5; Arizona, USA*

Photography by Jim Hughes, Assisted by ♪ Melody ♫
Collection of ♪ Melody ♫, Applewood, CO, USA

ENARGITE - Cu_3AsS_4 ;
Metallic greyish-black, grey/rose brown in reflected light; Hardness 3; Locality: Montana, USA

Photography by Jim Hughes, Assisted by ♪ Melody ♫
Collection of ♪ Melody ♫, Applewood, CO, USA

ENDLICHITE (On Descloizite)-
$(PbCl)Pb_4(VO_4)_3$;
Sub-translucent to opaque red, brown, yellow; Hardness 2.75-3; Locality: Nevada, USA

Photography by Jim Hughes, Assisted by ♪ Melody ♫
Collection of ♪ Melody ♫, Applewood, CO, USA

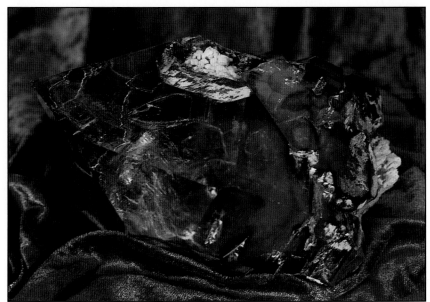

***ENHYDRO - (Quartz) - Mineral containing fluid deposited from that which
chalcedony, agate, or quartz was deposited; Hardness variable, but usually
silicate hardness of 7; Locality: Minas Gerais, Brasil
Photography by Jim Hughes, Assisted by ♪ Melody ♫
Collection of Bob Jackson, Applewood, CO, USA***

***ENHYDRO - (Calcite) - Locality: Mexico; Gift of Angel Torrecillas, USA
Photography by Jim Hughes, Assisted by ♪ Melody ♫
Collection of ♪ Melody ♫, Applewood, CO, USA***

ENSTATITE - *MgSiO₃* - *Pearly vitreous colourless, <u>grey</u>, green, yellow, brown; Hardness 5-6; Locality: Maryland, USA*

Photography by Jim Hughes, Assisted by ♪ Melody ♫
Collection of ♪ Melody ♫, Applewood, CO, USA

EOSPHORITE - *(Mn,Fe)AlPO₄(OH)₂♥H₂O - Vitreous resinous pink, <u>peach</u>, rose-red; Hardness 5; Connecticut, USA*

Photography by Jim Hughes, Assisted by ♪ Melody ♫
Collection of ♪ Melody ♫, Applewood, CO, USA

EPIDIDYMITE - *NaBeSi₃O₇(OH) - Transparent colourless; Hardness 5.5; Locality: Quebec, Canada*

Photography by Jim Hughes, Assisted by ♪ Melody ♫
Collection of ♪ Melody ♫, Applewood, CO, USA

EPIDOTE - *(With Quartz)*
$Ca_2FeAl_2(Si_2O_7)(SiO_4)(O,OH)_2$ -
*Vitreous yellowish-green, green,
brownish-green, black; Hardness
6; Locality: Prince of Wales,
Alaska, USA*

*Photography by Jim Hughes, Assisted by ♪ Melody ♫
Collection of Bob Jackson, Applewood, CO, USA*

EPISTILBITE -
$NaCa_3(Al_6Si_{18})O_{48}$♥$16H_2O$ -
*Vitreous colourless, white,
yellowish; Hardness 4-4.5;
Locality: Berufjord, Iceland*

*Photography by Jim Hughes, Assisted by ♪ Melody ♫
Collection of ♪ Melody ♫, Applewood, CO, USA*

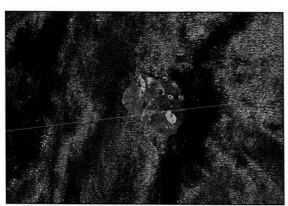

ERIONITE -
$K_2NaCa_{1.5}Mg(Al_8Si_{28})O_{72}$♥$28H_2O$ -
*Translucent whiteish; Hardness
from 5-5.5; Locality: Oregon,
USA*

*Photography by Jim Hughes, Assisted by ♪ Melody ♫
Collection of ♪ Melody ♫, Applewood, CO, USA*

ERYTHRITE -
$Co_3(AsO_4)_2 \cdot 8H_2O$ - *Adamantine pearly earthy red, pink; Hardness 1.5-2.5; Locality: Saxony, Germany*

Photography by Jim Hughes, Assisted by ♪ Melody ♫
Collection of ♪ Melody ♫, Applewood, CO, USA

ERYTHROSIDERITE -
$K_2FeCl_5 \cdot H_2O$ - *Vitreous red to orange/red; Hardness 2.5; Locality: Ural Mts., Russia*

Photography by Jim Hughes, Assisted by ♪ Melody ♫
Collection of ♪ Melody ♫, Applewood, CO, USA

ETTRINGITE -
$Ca_6Al_2(SO_4)_3(OH)_{12} \cdot 26H_2O$ - *Transparent colourless; Hardness 2-2.5; Locality: Republic of South Africa, Africa*

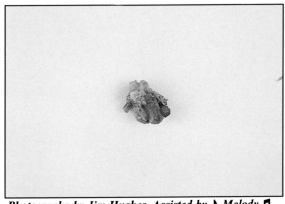

Photography by Jim Hughes, Assisted by ♪ Melody ♫
Collection of ♪ Melody ♫, Applewood, CO, USA

EUCHROITE - (Crystals) - $Cu_2AsO_4(OH)$♥$3H_2O$ - Vitreous Green; Hardness 3.5-4; Locality: Czechoslovakia

Photography by Jim Hughes, Assisted by ♪ Melody ♫
Collection of ♪ Melody ♫, Applewood, CO, USA

EUCLASE - $BeAlSiO_4(OH)$ - Vitreous pale green, blue; Hardness 7.5; Localilty: Republic of South Africa, Africa

Photography by Jim Hughes, Assisted by ♪ Melody ♫
Collection of ♪ Melody ♫, Applewood, CO, USA

EVANSITE - $Al_3PO_4(OH)_6$♥$6H_2O+$ - Vitreous resinous waxy colourless, white, various tints; Hardness 3-4; Locality: Barcelona, Spain

Photography by Jim Hughes, Assisted by ♪ Melody ♫
Collection of ♪ Melody ♫, Applewood, CO, USA

Within your love all time is bound ♥

[Bob Jackson]

FAUJASITE - *(Octohedral, with Phillipsite in Vugs) -*
$Na_{20}Ca_{12}Mg_8(Al_{60}Si_{132})O_{384}$♥
$235H_2O$ - *Vitreous adamantine colourless, white; Hardness 5; Locality: Kaiserstuhl, Germany*

Photography by Jim Hughes, Assisted by ♪ Melody ♫
Collection of ♪ Melody ♫, Applewood, CO, USA

FAUSTITE -
$(Zn,Cu)Al_6(PO_4)_4(OH)_8$♥$4H_2O;$
Waxy dull green; Hardness 5.5; Locality: Nevada, USA

Photography by Jim Hughes, Assisted by ♪ Melody ♫
Collection of ♪ Melody ♫, Applewood, CO, USA

FELDSPAR -
$(K,Na,Ca,Ba,NH_4)(Si,Al)_4O_8$ -
[General silicate] White to pale shades of yellow, red, green - sometimes dark; Hardness from 6-6.5; Locality: Mexico

Photography by Jim Hughes, Assisted by ♪ Melody ♫
Collection of ♪ Melody ♫, Applewood, CO, USA

FERGUSONITE - *YNbO₄* - *Vitreous sub-metallic black, greyish black; Hardness 5.5-6; Locality: Arendal, Norway*

FERGUSONITE - $YNbO_4$ -
Vitreous sub-metallic black,
greyish black; Hardness 5.5-6;
Locality: Arendal, Norway

Photography by Jim Hughes, Assisted by ♪ Melody ♫
Collection of ♪ Melody ♫, Applewood, CO, USA

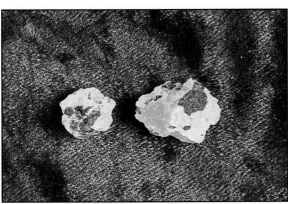

FERRIERITE -
$(Mg,K,Ca)_{4.4}(Si,Al)_{36}O_{72} \heartsuit 18H_2O$;
Whitish; Hardness 3; Locality:
British Columbia, Canada

Photography by Jim Hughes, Assisted by ♪ Melody ♫
Collection of ♪ Melody ♫, Applewood, CO, USA

FERSMANNITE -
$(Ca,Na)_8(Ti,Nb)_4(Si_2O_7)O_8F_3$ -
Vitreous dark brown to <u>red-brown</u>,
golden yellow; Hardness; 5-5.5;
Locality: Kola Peninsula, Russia

Photography by Jim Hughes, Assisted by ♪ Melody ♫
Collection of ♪ Melody ♫, Applewood, CO, USA

FIEDLERITE - *(Crystal in cavity)* - $Pb_3Cl_4(OH)_2$ - *Adamantine colourless, white; Hardness 3.5; Locality: Laurium, Greece.*

Photography by Jim Hughes, Assisted by ♪ Melody ♫
Collection of ♪ Melody ♫, Applewood, CO, USA

FILLOWITE - *(With Heterosite)* $Na_2Ca(Mn,Fe)_7(PO_4)_6$ - *Sub-resinous lustrous yellow, <u>yellowish-brown</u>, reddish-brown; Hardness 4.5; Locality: Rwanda (Kiluli), Africa*

Photography by Jim Hughes, Assisted by ♪ Melody ♫
Collection of ♪ Melody ♫, Applewood, CO, USA

FLINT - SiO_2 *with carbonaceous impurities; Sub-vitreous opaque greys, browns, blacks; Hardness 7; Locality: Ohio, USA*

Photography by Jim Hughes, Assisted by ♪ Melody ♫
Collection of Bob Jackson, Applewood, CO, USA

FLORENCITE - *(With Churchite); $3Al_2O_3 \heartsuit Ce_2O_3 \heartsuit 2P_2O_5 \heartsuit 6H_2O$ - Resinous to adamantine pale yellows; Hardness 5; Locality: California, USA*

Photography by Jim Hughes, Assisted by ♪ Melody ♫
Collection of ♪ Melody ♫, Applewood, CO, USA

FLUORITE - *CaF_2 - Vitreous colourless, green, blue, yellow, purple, pink, etc.; Hardness 4; Locality: Mexico*

Photography by Jim Hughes, Assisted by ♪ Melody ♫
Collection of ♪ Melody ♫, Applewood, CO, USA

FLUORITE - *(Octahedra); CaF_2 ; - Vitreous colourless, green, blue, yellow, purple, pink, etc.; Hardness 4; Locality: Illinois, USA*

Photography by Jim Hughes, Assisted by ♪ Melody ♫
Collection of ♪ Melody ♫, Applewood, CO, USA

FLUORITE - (Octahedra); CaF$_2$; - Vitreous colourless, green, blue, yellow,
*   purple, pink, etc.; Hardness 4; Locality: Mexico*
Photography by Jim Hughes, Assisted by ♪ Melody ♫
Collection of ♪ Melody ♫, Applewood, CO, USA

FLUORITE - *(Blue, with purple Calcite; CaF$_2$; - Vitreous colourless, green, blue, yellow, purple, pink, etc.; Hardness 4; Locality: Mexico*

Photography by Jim Hughes, Assisted by ♪ Melody ♫
Collection of ♪ Melody ♫, Applewood, CO, USA

FLUORITE - *(Green); CaF$_2$; - Vitreous colourless, green, blue, yellow, purple, pink, etc.; Hardness 4; Locality: Minas Gerais, Brasil*

Photography by Jim Hughes, Assisted by ♪ Melody ♫
Collection of ♪ Melody ♫, Applewood, CO, USA

FLUORITE - *(Golden-Brown); CaF$_2$; - Vitreous colourless, green, blue, yellow, purple, pink, etc.; Hardness 4; Locality: Minas Gerais, Brasil*

Photography by Jim Hughes, Assisted by ♪ Melody ♫
Collection of Julianne Guilbault, Lakewood, CO, USA

FLUORITE - *(Pink); CaF$_2$;*
Vitreous colourless, green, blue,
yellow, purple, pink, etc.;
Hardness 4; Locality: Mexico

Photography by Jim Hughes, Assisted by ♪ Melody ♫
Collection of ♪ Melody ♫, Applewood, CO, USA
Gift of Angel Torrecillas

FLUORITE - *(Pink); CaF$_2$;*
Vitreous colourless, green, blue,
yellow, purple, pink, etc.;
Hardness 4; Locality: Mexico

Photography by Jim Hughes, Assisted by ♪ Melody ♫
Collection of ♪ Melody ♫, Applewood, CO, USA
Gift from John Gallant, USA

FLUORITE - *(Yttrium);*
CaF$_2$ with varying amounts of
YF$_3$; - Vitreous colourless,
green, blue, yellow, purple, pink,
etc.; Hardness 4; Locality:
Mexico

Photography by Jim Hughes, Assisted by ♪ Melody ♫
Collection of ♪ Melody ♫, Applewood, CO, USA

FLUORITE - *(Chinese); CaF_2 ; Vitreous colourless, green, blue, yellow, purple, pink, etc., with pyrite; Hardness 4; Locality: China*

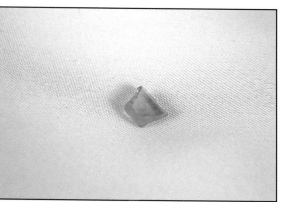

FORSTERITE - *Mg_2SiO_4 ; Translucent green, yellow; Hardness 7; Locality: Egypt*

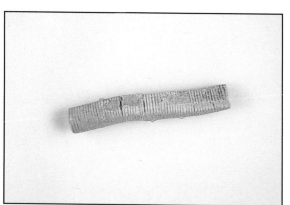

FOSSIL - *(Tube); Organic matter transformed to stone; Hardness variable dependent upon mineral composition; Locality: Oregon, USA*

FOSSIL - *(Leaf); Organic matter transformed to stone; Hardness variable dependent upon mineral composition; Locality: Utah, USA*

Photography by Jim Hughes, Assisted by ♪ Melody ♫
Collection of ♪ Melody ♫, Applewood, CO, USA

FOSSIL - *(Conglomerate); Organic matter transformed to stone; Hardness variable dependent upon mineral composition; Locality: Minnesota, USA*

Photography by Jim Hughes, Assisted by ♪ Melody ♫
Collection of Bob Jackson, Applewood, CO, USA

FOURMARIERITE - $PbO_3(UO_2)(OH)_4$♥$4H_2O$ - *Translucent reddish-orange, brown; Hardness 3-4; Locality: Bavaria, Germany*

Photography by Jim Hughes, Assisted by ♪ Melody ♫
Collection of ♪ Melody ♫, Applewood, CO, USA

FRANKLINITE -
$(Zn,Mn,Fe)(Fe,Mn)_2O_4$ -
*Metallic/sub-metallic black, white
in reflected light; Hardness 5.5-
6.5; Locality: New Jersey, USA*

Photography by Jim Hughes, Assisted by ♪ Melody ♫
Collection of ♪ Melody ♫, Applewood, CO, USA

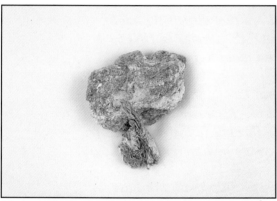

FUCHSITE: $(H,K)AlSiO_4$ *with
Chromium; Vitreous to silky
green; Hardness 2-2.5; Locality:
Minas Gerais, Brasil*

Photography by Jim Hughes, Assisted by ♪ Melody ♫
Collection of ♪ Melody ♫, Applewood, CO, USA

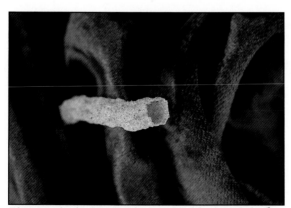

FULGURITE - *[Natural Glass
(Lechatelierite)]; Fused SiO_2;
Vitreous colourless, sandy colour;
Hardness 7, but structure brittle;
Locality: Ocean beach, Republic
Of South Africa, Africa*

Photography by Jim Hughes, Assisted by ♪ Melody ♫
Collection of ♪ Melody ♫, Applewood, CO, USA

Breathe

Breathe deep the gathering Light,
Energy of Love that dispels the night.
ALL THAT IS vibrates as ONE,
From tiniest atom to mightiest sun.

From ephemeral lace to the galaxy's core -
Power of Love - no less, always more,
Growing while flowing in the river of Life,
Harmonics converge and merge in the Light.

Breathe deep - filled with the Love
Of the ALL that's below and the All that's above.
ALL THAT IS lives inside of you,
Ancient of age and forever brand new.

[Richard Ray, Washington, USA]

GALENA - *PbS - Metallic grey, white in reflected light; Hardness 2.6; Locality: Missour, USA*

Photography by Jim Hughes, Assisted by ♪ Melody ♫
Collection of ♪ Melody ♫, Applewood, CO, USA

GANOPHYLLITE - $(K,Na)_6(Mn,Al,Mg)_{24}(Si,Al)_{40}O_{96}(OH)_{16}$♥$21H_2O$ - *Translucent brown; Hardness 4-4.5; Locality: New Jersey, USA*

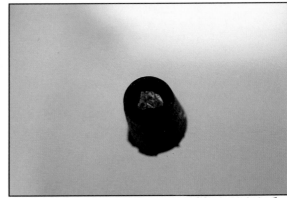

Photography by Jim Hughes, Assisted by ♪ Melody ♫
Collection of Colorado School Of Mines, Colorado, USA

GARNET -$(Ca,Fe,Mg,Mn)_3(Al,Fe,Mn,Cr,Ti,V)_2(SiO_4)_3$ - *[General orthosilicate] - Hardness 6.5-7.5; Locality: Zimbabwe, Africa*

Photography by Jim Hughes, Assisted by ♪ Melody ♫
Collection of ♪ Melody ♫, Applewood, Colorado, USA
Gift from Bob Jackson, Applewood, Colorado, USA

GARNET - *(In Schist);*
(Ca,Fe,Mg,Mn)$_3$
(Al,Fe,Mn,Cr,Ti,V)$_2$(SiO$_4$)$_3$ -
[General orthosilicate] - Hardness
6.5-7.5; Locality: Idaho, USA

Photography by Jim Hughes, Assisted by ♪ Melody ♫
Collection of ♪ Melody ♫, Applewood, CO, USA

GAUDEFROYITE - *(Crystals in*
Manganite);
Ca$_4$Mn$_{3-x}$(BO$_3$)$_3$(CO$_3$)(O,OH)$_3$;
Brilliant luster to dull black;
Hardness 6; Locality: Buranga,
Rwanda, Africa

Photography by Jim Hughes, Assisted by ♪ Melody ♫
Collection of ♪ Melody ♫, Applewood, CO, USA

GAUDEFROYITE - *(Crystal);*
Ca$_4$Mn$_{3-x}$(BO$_3$)$_3$(CO$_3$)(O,OH)$_3$;
Brilliant luster to dull black;
Hardness 6; Locality: Buranga,
Rwanda, Africa

Photography by Jim Hughes, Assisted by ♪ Melody ♫
Collection of ♪ Melody ♫, Applewood, CO, USA

GEHLENITE - $Ca_2Al(Si,Al)_2O_7$ - *Translucent colourless, white, <u>grey</u>, greyish-green, brown; Hardness 5-6; Locality: California, USA.*

Photography by Jim Hughes, Assisted by ♪ Melody ♫
Collection of ♪ Melody ♫, Applewood, CO, USA

GENTHELVITE - *(In Hematite)* $Be_3Zn_4(SiO_4)_3S$ - *Translucent green, <u>bluish</u>, red, yellow; Hardness 6-6.5; Locality: Argentina, South America*

Photography by Jim Hughes, Assisted by ♪ Melody ♫
Collection of ♪ Melody ♫, Applewood, CO, USA

GEODE - *Spherical configuration containing a cavity lined with crystalline structures growing toward the center. These crystalline structures occur in square configuration - composition of crystalline structure not known at this time.*

Photography by Jim Hughes, Assisted by ♪ Melody ♫
Collection of Howard Dolph, Oregon, USA

GEODE - Spherical configuration containing a cavity lined with crystalline structures growing toward the center. These crystalline structures occur in square configuration.

Photography by Jim Hughes, Assisted by ♪ Melody ♫
Collection of Howard Dolph, Oregon, USA

GIBBSITE - *gamma-Al(OH)$_3$ -
Vitreous pearly white, greyish,
greenish; Hardness 2.5-3.5;
Locality: Telemark, Norway*

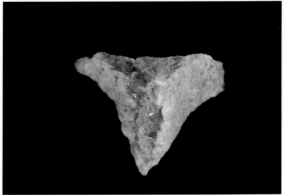

Photography by Jim Hughes, Assisted by ♪ Melody ♫
Collection of ♪ Melody ♫, Applewood, CO, USA

GILSONITE - *Black brilliant
lustrous; Oxygenated
hydrocarbonaceous asphalt;
Hardness 2-2.5; Locality: Utah,
USA*

Photography by Jim Hughes, Assisted by ♪ Melody ♫
Collection of ♪ Melody ♫, Applewood, CO, USA

GISMONDINE - *(On
Phillipsite) Ca$_2$Al$_4$Si$_4$O$_{16}$♥9H$_2$O -
Vitreous colourless, <u>whitish</u>,
greyish; Hardness 4.5; Locality:
Styria, Australia*

Photography by Jim Hughes, Assisted by ♪ Melody ♫
Collection of ♪ Melody ♫, Applewood, CO, USA

GMELINITE -
$Na_4(Al_4Si_8)O_{24} \heartsuit 11H_2O$ - *Vitreous*
colourless, *yellowish-white,*
greenish-white, reddish; Hardness
4.5; Locality: Quebec, Canada

GOETHITE - *alpha-FeO(OH)* -
Adamantine-metallic, earthy-
brown, yellow, black, grey;
Hardness 5-5.5; Locality:
Czechoslovakia

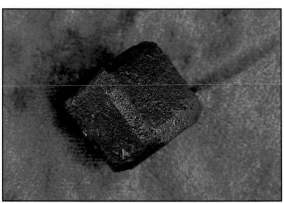

GOETHITE - *alpha-FeO(OH)* -
Adamantine-metallic, earthy-
brown, yellow, black, grey;
Hardness 5-5.5; Locality: South
Australia, Australia

GOETHITE - (With Quartz);
alpha-FeO(OH); Adamantine-
metallic, earthy-brown, yellow,
black, grey; Hardness 5-5.5;
Locality: Minas Gerais, Brasil

Photography by Jim Hughes, Assisted by ♪ Melody ♫
Collection of ♪ Melody ♫, Applewood, CO, USA

GOETHITE - (With Amethyst);
alpha-FeO(OH); Adamantine-
metallic, earthy-brown, yellow,
black, grey; Hardness 5-5.5;
Locality: Minas Gerais, Brasil

Photography by Jim Hughes, Assisted by ♪ Melody ♫
Collection of ♪ Melody ♫, Applewood, CO, USA

GOLD - (Crystal); Au; Metallic
yellow; Hardness 2.5-3;
Locality: California, USA

Photography by Jim Hughes, Assisted by ♪ Melody ♫
Collection of Bob Jackson, Applewood, CO, USA

GOLD - *(Native, in Quartz); Au;*
Metallic yellow; Hardness 2.5-3;
Locality: Quebec, Canada

Photography by Jim Hughes, Assisted by ♪ Melody ♫
Collection of Bob Jackson, Applewood, CO, USA

GOLD - *(In Quartz Crystal); Au;*
Metallic yellow; Hardness 2.5-3;
Locality: Western USA

Photography by Jim Hughes, Assisted by ♪ Melody ♫
Collection of ♪ Melody ♫, Applewood, CO, USA

GOSHENITE - *(With Mica);*
$Be_3Al_2(SiO_3)_6$ - *Transparent to*
translucent with vitreous luster
colourless; Hardness 7.5-8;
Locality: Minas Gerais, Brasil

Photography by Jim Hughes, Assisted by ♪ Melody ♫
Collection of ♪ Melody ♫, Applewood, CO, USA

GOSHENITE - *(With Tourmaline);*
$Be_3Al_2(SiO_3)_6$ - *Transparent to translucent with vitreous luster colourless; Hardness 7.5-8; Locality: Minas Gerais, Brasil*

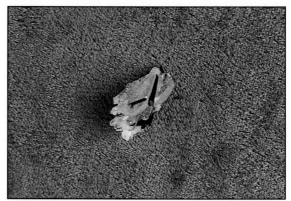

Photography by Jim Hughes, Assisted by ♪ Melody ♫
Collection of ♪ Melody ♫, Applewood, CO, USA

GOYAZITE - *(Yellow crystals in Light Blue Fluorapatite);*
$SrAl_3(PO_4)(PO_3OH)(OH)_6$;
Lustrous resinous pink, <u>yellow</u>; Hardness 4.5-5; Locality: South Dakota, USA

Photography by Jim Hughes, Assisted by ♪ Melody ♫
Collection of ♪ Melody ♫, Applewood, CO, USA

GRANDIDIERITE -
$(Mg,Fe)Al_3BSiO_9$ - *Translucent bluish-green; Hardness 7.5; Locality:Madagascar*

Photography by Jim Hughes, Assisted by ♪ Melody ♫
Collection of ♪ Melody ♫, Applewood, CO, USA

GRANITE - (With Tourmaline); Medium to coarse-grained plutonic rock consisting chiefly of quartz and feldspar, orthoclase, and/or mica and hornblende - occasionally with tourmaline; Hardness variable; Locality: Cruceiro Do Sol, Rio, Brasil

Photography by ♪ Melody ♫
Cruceiro Do Sol, Minas Gerais, Brasil

GRANITE - Medium to coarse-grained plutonic rock consisting chiefly of quartz and feldspar, orthoclase, and/or mica and hornblende; Hardness variable; Locality: Ouro Preto, Minas Gerais, Brasil

Photography by ♪ Melody ♫
Ouro Preto, Minas Gerai, Brasil

GRANITE - Medium to coarse-grained plutonic rock consisting chiefly of quartz and feldspar, orthoclase, and/or mica and hornblende; Hardness variable; Locality: India

Photography by Jim Hughes, Assisted by ♪ Melody ♫
Collection of Bob Jackson, Applewood, CO, USA

GRAPHITE - *C* - *Metallic black/grey; Hardness 1-2; Locality: Ratnapura, Sri Lanka*

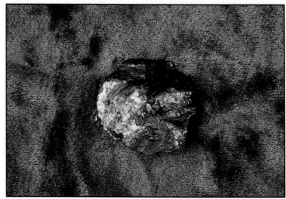

GREENOCKITE - *(With Sphalerite and Galena); beta-CdS; Adamantine resinous yellow, orange; Hardness 3-3.5; Locality: Connecticut,USA*

GROSSULAR GARNET - $Ca_3Al_2(SiO_4)_3$ - *Vitreous to resinous browns, <u>greens</u>, yellows, red-brown, reds, oranges, colourless, white, grey, and black; Hardness 6.5-7.5; Locality: Siberia, Russia*

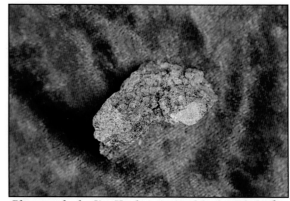

GROSSULAR GARNET -
(Transvaal Jade) - $Ca_3Al_2(SiO_4)_3$ -
Vitreous to resinous browns,
greens, yellows, red-brown, reds,
oranges, colourless, white, grey,
and black; Specifically opaque to
translucent massive grossular
garnet green in colour, sometimes
occurring with reddish translucent
grossular garnet; Hardness 6.5-
7.5; Locality: Republic of South
Africa, Africa

Photography by Jim Hughes, Assisted by ♪ Melody ♫
Collection of Bob Jackson, Applewood, CO, USA

GYPSUM - *(Crystal)* -
$CaSO_4 ♥ 2H_2O$ *- Sub-vitreous*
<u>*colourless, white*</u>*, grey, yellowish,*
brownish; Hardness 2; Locality:
California, USA

Photography by Jim Hughes, Assisted by ♪ Melody ♫
Collection of Bob Jackson, Applewood, CO, USA

GYPSUM - *(Red with Red*
***Quartz)*;** $CaSO_4 ♥ 2H_2O$ *- Sub-*
vitreous; Hardness 2; Locality:
Teruel, Spain

Photography by Jim Hughes, Assisted by ♪ Melody ♫
Collection of ♪ Melody ♫, Applewood, CO, USA

Unconditional love sets the spirit free

[Julianne Guilbault]

HALITE - *NaCl; Vitreous <u>pink</u>, colourless, white, yellow, red, blue, purple; Hardness 2; Locality: California, USA*

Photography by Jim Hughes, Assisted by ♪ Melody ♫
Collection of Jackson & Melody, Applewood, CO, USA
Gift of W.R. Horning, California, USA

HALITE - *NaCl; Vitreous pink, colourless, white, yellow, red, blue, <u>purple</u>; Hardness 2; Locality: Mulhouse, France*

Photography by Jim Hughes, Assisted by ♪ Melody ♫
Collection of ♪ Melody ♫, Applewood, CO, USA

HALITE - *NaCl; Vitreous pink, colourless, <u>white</u>, yellow, red, blue, purple; Hardness 2; Locality: Utah, USA*

Photography by Jim Hughes, Assisted by ♪ Melody ♫
Collection of ♪ Melody ♫, Applewood, CO, USA

HAMBERGITE - *(With Matrix) - $Be_2BO_3(OH)$ - Vitreous colourless, greyish, <u>white, yellowish</u>; Hardness 7.5; Locality: California, USA*

Photography by Jim Hughes, Assisted by ♪ Melody ♫
Collection of ♪ Melody ♫, Applewood, CO, USA

HANKSITE - *$KNa_{22}(SO_4)_9(CO_3)_2Cl$; Vitreous/dull colourless, yellowish-green, grey; Hardness 3-3.5; Locality: California, USA*

Photography by Jim Hughes, Assisted by ♪ Melody ♫
Collection of ♪ Melody ♫, Applewood, CO, USA
Gift of W.R. Horning, California, USA

HARKERITE - *$Ca_{12}Mg_4Al(CO_3)_4(BO_3)_4$ $(SiO_4)_4(H_2O,Cl)_{17}$ - Vitreous colourless to grey/white; Hardness ?; Isle of Skye, Scotland*

Photography by Jim Hughes, Assisted by ♪ Melody ♫
Collection of ♪ Melody ♫, Applewood, CO, USA

HARMOTOME -
$Ba_2(Ca_{0.5}Na)(Si_{11}Al_5)O_{32} \heartsuit 12H_2O$
- *Vitreous colourless, white,*
grey, yellow, red, brown;
Hardness 4.5; Locality: Scotland

Photography by Jim Hughes, Assisted by ♪ Melody ♫
Collection of ♪ Melody ♫, Applewood, CO, USA

HAZELWOODITE - *(Metallic,*
in Serpentine with green
Zaratite); Hardness ?; Locality:
Tasmania, Australia

Photography by Jim Hughes, Assisted by ♪ Melody ♫
Collection of ♪ Melody ♫, Applewood, CO, USA

HEINRICHITE -
$Ba(UO_2)_2(AsO_4)_2 \heartsuit 10H_2O$ -
Vitreous pearly yellow to rust,
green; Hardness 2.5; Locality:
Black Forest, Germany

Photography by Jim Hughes, Assisted by ♪ Melody ♫
Collection of ♪ Melody ♫, Applewood, CO, USA

HELIODOR - $Be_3Al_2Si_6O_{18}$ -
Vitreous gold; Hardness 7.5-8;
Locality: Minas Gerais, Brasil

Photography by Jim Hughes, Assisted by ♪ Melody ♫
Collection of ♪ Melody ♫, Applewood, CO, USA

HEMATITE *(Botryoidal); alpha-*
Fe_2O_3 - *Metallic/earthy grey, red;*
Hardness 5-6; Locality:
Minnesota, USA

Photography by Jim Hughes, Assisted by ♪ Melody ♫
Collection of ♪ Melody ♫, Applewood, CO, USA

HEMATITE *(Mountain); alpha-*
Fe_2O_3 - *Metallic/earthy grey, red;*
Hardness 5-6; Locality: "Top Of
The World", Belo Horizonte
(near), Minas Gerais, Brasil

Photography by ♪ Melody ♫

HEMATITE (Red Crystals);
alpha-Fe$_2$O$_3$ - Metallic/earthy
grey, red; Hardness 5-6;
Locality: Argentina, South
America

Photography by Jim Hughes, Assisted by ♪ Melody ♫
Collection of ♪ Melody ♫, Applewood, CO, USA

HEMATITE (Key Crystal);
alpha-Fe$_2$O$_3$ - Metallic/earthy
grey, red; Hardness 5-6;
Locality: Minas Gerais, Brasil

Photography by Jim Hughes, Assisted by ♪ Melody ♫
Collection of Bob Jackson, Applewood, CO, USA

HEMATITE (Rainbow); alpha-
Fe$_2$O$_3$ - Metallic/earthy grey,
red; Hardness 5-6; Locality:
Minas Gerais, Brasil

Photography by Jim Hughes, Assisted by ♪ Melody ♫
Collection of ♪ Melody ♫, Applewood, CO, USA

HEMATITE *(With Quartz);*
alpha-Fe$_2$O$_3$ - Metallic/earthy
grey, red; Hardness 5-6; Locality:
Minas Gerais, Brasil

Photography by Jim Hughes, Assisted by ♪ Melody ♫
Collection of ♪ Melody ♫, Applewood, CO, USA

HEMIMORPHITE -
Zn$_4$Si$_2$O$_7$(OH)$_2$♥H$_2$O - Vitreous
pearly white, <u>bluish, greenish,</u>
yellowish, brown; Hardness 4.5-5;
Locality: Mexico

Photography by Jim Hughes, Assisted by ♪ Melody ♫
Collection of ♪ Melody ♫, Applewood, CO, USA

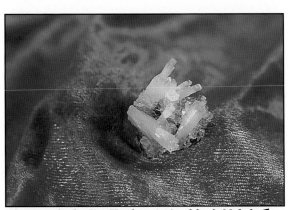

HEMIMORPHITE -
Zn$_4$Si$_2$O$_7$(OH)$_2$♥H$_2$O - Vitreous
pearly <u>white</u>, bluish, greenish,
yellowish, brown; Hardness 4.5-5;
Locality: Mexico

Photography by Jim Hughes, Assisted by ♪ Melody ♫
Collection of ♪ Melody ♫, Applewood, CO, USA

HEMIMORPHITE -
$Zn_4Si_2O_7(OH)_2 \cdot H_2O$ - *Vitreous*
pearly white, bluish, greenish,
yellowish, brown; Hardness
from 4.5-5; Locality: Mexico

Photography by Jim Hughes, Assisted by ♪ Melody ♫
Collection of ♪ Melody ♫, Applewood, CO, USA

HERDERITE - $CaBePO_4(F,OH)$
- Vitreous colourless, pale
yellow, greenish-white;
Hardness 5-5.5; Locality: Minas
Gerais, Brasil

Photography by Jim Hughes, Assisted by ♪ Melody ♫
Collection of ♪ Melody ♫, Applewood, CO, USA

HERKIMER DIAMOND - SiO_2;
- Crystalline quartz configured
as double-terminated short,stout,
clear or included prismatic
crystals; Hardness 7; Locality:
New York, USA

Photography by Jim Hughes, Assisted by ♪ Melody ♫
Collection of ♪ Melody ♫, Applewood, CO, USA

HERKIMER DIAMONDS - SiO_2; - *Crystalline quartz (Above, Amethyst [Gift of LaSonda Sioux Sipe, Virginia, USA]; Locality Tanzania, Africa) (Below, Clear Quartz; Location New York, USA) configured as double- terminated short,stout, clear or included prismatic crystals; Hardness 7. Photography by Jim Hughes, Assisted by ♪ Melody ♫. Collection of ♪ Melody ♫.*

HESSONITE - $Ca_3Al_2(SiO_4)_3$ - Vitreous to resinous cinnamon-to-yellow colour; Hardness 6.5-7.5; Locality: Republic of South Africa, Africa

Photography by Jim Hughes, Assisted by ♪ Melody ♫
Collection of ♪ Melody ♫, Applewood, CO, USA

HEULANDITE - $(Na,K,Ca,Sr,Ba)_5(Al_9Si_{27})O_{72}$♥2 $6H_2O$ - Vitreous pearly white, red, peach, grey, brown; Hardness 3.5-4; Locality: Maharashtra, India

Photography by Jim Hughes, Assisted by ♪ Melody ♫
Collection of ♪ Melody ♫, Applewood, CO, USA

HIDDENITE - $LiAl(SiO_3)_2$ - Vitreous yellow-green to green colour; Hardness 6.5-7; Locality: Minas Gerais, Brasil

Photography by Jim Hughes, Assisted by ♪ Melody ♫
Collection of ♪ Melody ♫, Applewood, CO, USA

HIDDENITE - *LiAl(SiO₃)₂* - *(Faceted); Vitreous yellow-green to green colour; Hardness 6.5-7; Locality: Minas Gerais, Brasil*

Photography by Jim Hughes, Assisted by ♪ Melody ♫
Collection of ♪ Melody ♫, Applewood, CO, USA
Gift of Gustavo Ferraz Oliveira, Belo Horizonte,
Minas Gerais, Brasil

HODGKINSONITE - *(Purple, with black Franklinite & Willemite); Zn₂MnSiO₄(OH)₂ - Translucent reddish/violet pink, reddish-brown, orange; Hardness 4.5-5; Locality: New Jersey, USA*

Photography by Jim Hughes, Assisted by ♪ Melody ♫
Collection of ♪ Melody ♫, Applewood, CO, USA

HOMILITE - *Ca₂(Fe,Mg)B₂Si₂O₁₀ - Resinous/vitreous black, blackish-brown; Hardness 5; Locality: Langesundfjord, Norway*

Photography by Jim Hughes, Assisted by ♪ Melody ♫
Collection of ♪ Melody ♫, Applewood, CO, USA

HOPEITE - $Zn_3(PO_4)_2 \cdot 4H_2O$ - Vitreous colourless, greyish-white, pale yellow, <u>orange</u>; Hardness 3.2; Locality: Zimbabwe, Africa

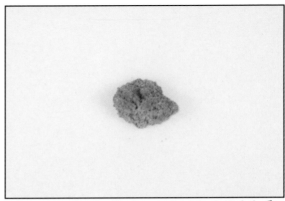

Photography by Jim Hughes, Assisted by ♪ Melody ♫
Collection of ♪ Melody ♫, Applewood, CO, USA

HORNEBLENDE - *(Red-brown "hairs")* - $Ca_2(Mg,Fe,Al)_5(Si,Al)_8O_{22}(OH)_2$ - *[General amphibole] Colour between black and white, greens, dark brown, yellow, pink, rose-red; Hardness 5-6; Locality: East Germany*

Photography by Jim Hughes, Assisted by ♪ Melody ♫
Collection of ♪ Melody ♫, Applewood, CO, USA

HOWLITE - $Ca_2B_5SiO_9(OH)_5$ - Sub-vitreous white; Hardness 3.5; Locality: California, USA

Photography by Jim Hughes, Assisted by ♪ Melody ♫
Collection of Bob Jackson, Applewood, CO, USA

HOWLITE - $Ca_2B_5SiO_9(OH)_5$ - *(Photo shows exterior coating); Hardness 3.5; Locality: California, USA*

Photography by Jim Hughes, Assisted by ♪ Melody ♫
Collection of Bob Jackson, Applewood, CO, USA

HUBNERITE *(Crystals with Pyrite in Quartz)* - $MnWO_4$; *Submetallic luster, grey to brown-black; Hardness 5-5.5; Locality: Montana, USA*

Photography by Jim Hughes, Assisted by ♪ Melody ♫
Collection of ♪ Melody ♫, Applewood, CO, USA

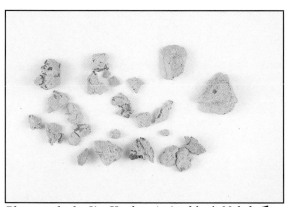

HUMMERITE - $KMgV_5O_{14}♥8H_2O$ - *Translucent orange, tan; Hardness ?; Locality: Colorado, USA*

Photography by Jim Hughes, Assisted by ♪ Melody ♫
Collection of ♪ Melody ♫, Applewood, CO, USA

HYDROZINCITE -
$Zn_5(CO_3)_2(OH)_6$ - *Earthy-silky white, grey, yellowish, browhish, pinkish, etc.; Hardness 2-2.5; Locality: Nevada, USA*

Photography by Jim Hughes, Assisted by ♪ Melody ♫
Collection of ♪ Melody ♫, Applewood, CO, USA

HYPERSTHENE - $(Fe,Mg)SiO_3$
- Dark brownish green, greyish-black, greenish-black, brown; Hardness 5-6; Locality: Canada

Photography by Jim Hughes, Assisted by ♪ Melody ♫
Collection of Bob Jackson, Applewood, CO, USA

HYPERSTHENE - $(Fe,Mg)SiO_3$
- (Eulite variety)- Dark brownish green, greyish-black, greenish-black, brown; Hardness 5-6; Locality: New York, USA

Photography by Dave Shrum, Colorado Camera Co., Lakewood, Colorado, USA
Collection of ♪ Melody ♫, Applewood, CO, USA

IDOCRASE -
$Ca_{19}Fe(Mg,Al)_8Al_4(SiO_4)_{10}$
$(Si_2O_7)_4(OH)_{10}$ -
Vitreous/resinous brown, <u>green,</u>
<u>yellow,</u> pale blue; Hardness 6.5;
Locality: Nevada, USA

Photography by Jim Hughes, Assisted by ♪ Melody ♫
Collection of ♪ Melody ♫, Applewood, CO, USA

ILMENITE - (Crystals with
Magnetite Crystals) - $FeTiO_3$;
Metallic/sub-metallic   black;
Hardness 5-6; Canada

Photography by Jim Hughes, Assisted by ♪ Melody ♫
Collection of ♪ Melody ♫, Applewood, CO, USA

ILVAITE - (With Calcite);
$CaFe_3O(Si_2O_7)(OH)$ - Sub-
metallic black, dark greyish-
black; Hardness 5.5-6; Locality:
Idaho, USA

Photography by Jim Hughes, Assisted by ♪ Melody ♫
Collection of ♪ Melody ♫, Applewood, CO, USA

INESITE -
$H_2(Mn,Ca)_6Si_6O_{19} \cdot 3H_2O$; *Pink crystalline; Hardness 6; Locality: California, USA*

Photography by Jim Hughes, Assisted by ♪ Melody ♫
Collection of ♪ Melody ♫, Applewood, CO, USA

INESITE -
$H_2(Mn,Ca)_6Si_6O_{19} \cdot 3H_2O$; *Red crystalline; Hardness 6; Locality: Republic of South Africa, Africa*

Photography by Jim Hughes, Assisted by ♪ Melody ♫
Collection of ♪ Melody ♫, Applewood, CO, USA

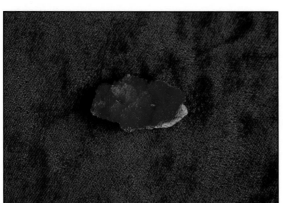

IOLITE - $Mg_2Al_4Si_5O_{18}$ -
(Faceted); Translucent greyish, lilac, blue with strong pleochroism; Hardness 7; Locality: Sri Lanka

Photography by Jim Hughes, Assisted by ♪ Melody ♫
Collection of Bob Jackson, Applewood, CO, USA

IRIDOSOMINE *- Composed of Iridium, Osmium, Rhodium, Platinum, Ruthenium, etc.; Metallic lustrous opaque tin-white to light steel-grey small flattened grains; Hardness 6-7; Locality: Ural Mountains, Russia*

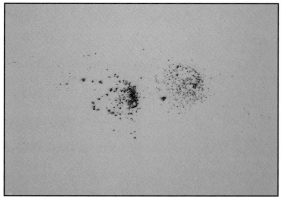

Photography by Jim Hughes, Assisted by ♪ Melody ♫
Collection of ♪ Melody ♫, Applewood, CO, USA

IRON *(Terrestrial) - alpha-Fe - Metallic grey; Hardness 4; Locality: Krasnonarski Krai, Russia*

Photography by Jim Hughes, Assisted by ♪ Melody ♫
Collection of ♪ Melody ♫, Applewood, CO, USA

IVORY *- (Tagua Nut - Vegetable Ivory) (shown); Hardness ?; Locality: Latin America*

Photography by Jim Hughes, Assisted by ♪ Melody ♫
Collection of Bob Jackson, Applewood, CO, USA

JADEITE - $Na(Al,Fe)Si_2O_6$ -
Vitreous colourless, white,
<u>greenish-grey</u>, greenish-blue,
blue, purple, etc.; Hardness 6;
Locality: Central America

Photography by Jim Hughes, Assisted by ♪ Melody ♫
Collection of ♪ Melody ♫, Applewood, CO, USA

JADEITE - $Na(Al,Fe)Si_2O_6$ -
Vitreous colourless, white,
greenish-grey, greenish-blue,
blue, purple, etc.; Hardness 6;
Locality: Burma

Photography by Jim Hughes, Assisted by ♪ Melody ♫
Collection of Bob Jackson, Applewood, CO, USA

JADEITE - $Na(Al,Fe)Si_2O_6$ -
Vitreous colourless, white,
greenish-grey, greenish-blue,
blue, <u>purple</u>, etc.; Hardness 6;
Locality: Middle East

Photography by Jim Hughes, Assisted by ♪ Melody ♫
Collection of Bob Jackson, Applewood, CO, USA

JAHNSITE -
$CaMn(Mg,Fe)_2Fe_2^{3+}(PO_4)_4(OH)_2 \cdot 8H_2O$ **- Vitreous/sub-adamantine brown, <u>yellow</u>, greenish-yellow, yellowish-brown, etc., Hardness 4; Locality: South Dakota, USA.**

Photography by Jim Hughes, Assisted by ♪ Melody ♫
Collection of ♪ Melody ♫, Applewood, CO, USA

JASPER - (Bat Cave); SiO_2 with impurities; Hardness 7; Locality: Oregon, USA

Photography by Jim Hughes, Assisted by ♪ Melody ♫
Collection of Bob Jackson, Applewood, CO, USA

JASPER - (Bruneau); SiO_2 with impurities; Hardness 7; Locality: Idaho/Oregon, USA

Photography by Jim Hughes, Assisted by ♪ Melody ♫
Collection of ♪ Melody ♫, Applewood, CO, USA
Gift from Bob Jackson, Applewood, CO, USA

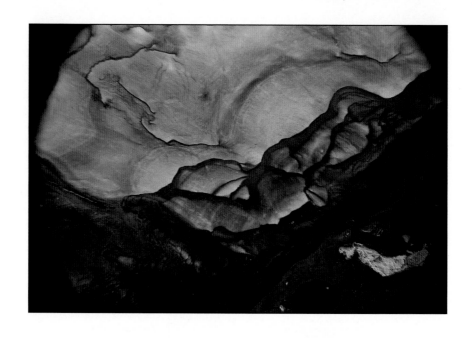

JASPER - (Biggs Blue Picture [Above] and Biggs Picture [Below]);
SiO$_2$ with impurities; Hardness 7; Locality: Oregon, USA
Photography by Jim Hughes, Assisted by ♪ Melody ♫
Collection of ♪ Melody ♫, Applewood, CO, USA

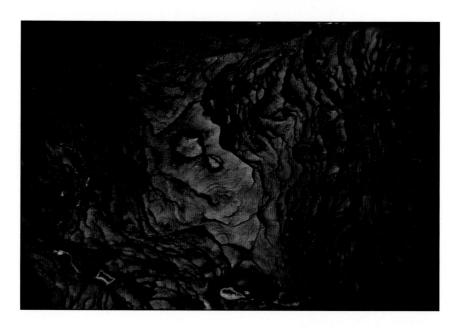

JASPER - (Orbicular); SiO₂ with impurities; Hardness 7; Locality: Washington, USA

Photography by Jim Hughes, Assisted by ♪ Melody ♫
Collection of Bob Jackson, Applewood, CO, USA

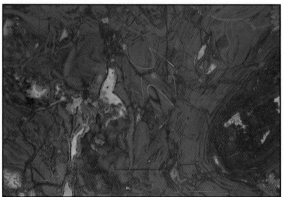

JASPER - (Red); SiO₂ with impurities; Hardness 7; Locality: Wyoming, USA

Photography by Jim Hughes, Assisted by ♪ Melody ♫
Collection of Bob Jackson, Applewood, CO, USA

JASPER - (Rose-Eye); SiO₂ with impurities; Hardness 7; Locality: Mexico

Photography by Jim Hughes, Assisted by ♪ Melody ♫
Collection of ♪ Melody ♫, Applewood, CO, USA

JASPER - *(Royal Plume); SiO_2 with impurities; Hardness 7; Locality: India*

Photography by Jim Hughes, Assisted by ♪ Melody ♫
Collection of Bob Jackson, Applewood, CO, USA

JASPER - *(Wonderstone); SiO_2 with impurities; Hardness 7; Locality: Utah, USA*

Photography by Jim Hughes, Assisted by ♪ Melody ♫
Collection of Bob Jackson, Applewood, CO, USA

JEREMEJEVITE - *$Al_6(BO_3)_5(F,OH)_3$ - Vitreous colourless,* <u>*pale yellowish-brown*</u>*; Hardness 7.5; Namibia, Africa*

Photography by Jim Hughes, Assisted by ♪ Melody ♫
Collection of ♪ Melody ♫, Applewood, CO, USA

JET - C_nH_{2n+2} - *Oxygenated hydrocarbon opaque black resinous to lustrous; Hardness 0.5-2.5; Locality: Whitby, England*

Photography by Jim Hughes, Assisted by ♪ Melody ♫
Collection of Bob Jackson, Applewood, CO, USA

JOAQUINITE - *(With Natrolite and Neptunite)* - $NaBa_2FeTi_2Ce_2(SiO_3)_8O_2(OH)$♥ H_2O - *Silky brown, yellowish brown, rusty brown; Hardness 5; Locality: California, USA*

Photography by Jim Hughes, Assisted by ♪ Melody ♫
Collection of ♪ Melody ♫, Applewood, CO, USA

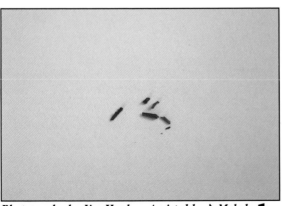

JULIENITE - $Na_2Co(SCN)_4$♥$8H_2O$ - *Translucent blue to black; Hardness 1.5-2; Locality: Shaba, Zaire*

Photography by Jim Hughes, Assisted by ♪ Melody ♫
Collection of ♪ Melody ♫, Applewood, CO, USA

As Long As You're Green, You'll Grow

[Alvin Davis, Columbus, Georgia, USA]

KAINOSITE -
$Ca_2(Y,Ce)_2(SiO_3)_4(CO_3)♥H_2O$ -
Lustrous <u>peach to pink</u>,
yellowish-brown; Hardness 5.5;
Locality; Ontario, Canada

Photography by Jim Hughes, Assisted by ♪ Melody ♫
Collection of Colorado School Of Mines, Colorado, USA

KAMMERERITE -
$H_8(Mg,Fe)_5Al_2Si_3O_{18}$ - Vitreous
to pearly red to peachy-red;
Hardness 2-2.5; Locality:
Erzurum, Turkey

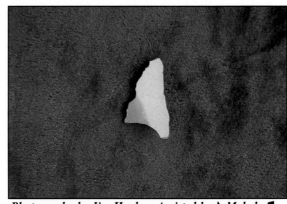

Photography by Jim Hughes, Assisted by ♪ Melody ♫
Collection of ♪ Melody ♫, Applewood, CO, USA

KAOLINITE - $Al_2Si_2O_5(OH)$ -
Pearly/earthy white, greyish,
yellowish, brownish, etc.;
Hardness 2-2.5; Locality:
Oregon, USA

Photography by Jim Hughes, Assisted by ♪ Melody ♫
Collection of ♪ Melody ♫, Applewood, CO, USA

KATOPHORITE - *(Black Crystals)* - $Na_2Ca(Mg,Fe)_5(Si,Al)_8O_{22}(OH)_2$ - *[General amphibole]; Colour ranges from* <u>*black*</u> *to reddish/ greenish, brown/yellow, greenish- blue; Hardness 5-6; Locality: Quebec, Canada*

Photography by Jim Hughes, Assisted by ♪ Melody ♫
Collection of ♪ Melody ♫, Applewood, CO, USA

KIDWELLITE - *Composition unknown to author; Locality: Arkansas, USA*

Photography by Jim Hughes, Assisted by ♪ Melody ♫
Collection of ♪ Melody ♫, Applewood, CO, USA

KINOITE - $Cu_2Ca_2Si_3O_{10}$♥$2H2_0$
Drusy deep blue crystalline silicates on quartzite; Locality: Arizona, USA

Photography by Jim Hughes, Assisted by ♪ Melody ♫
Collection of ♪ Melody ♫, Applewood, CO, USA
Gift of W.R. Horning, California, USA

KOLBECKITE - *(With green Overite); ScPO₄♥2H₂O; Vitreous/pearly blue, blue-grey; Hardness 3.5-4; Locality: Utah, USA*

KOLBECKITE - *(With green Overite);* $ScPO_4 \cdot 2H_2O$; *Vitreous/pearly blue, blue-grey; Hardness 3.5-4; Locality: Utah, USA*

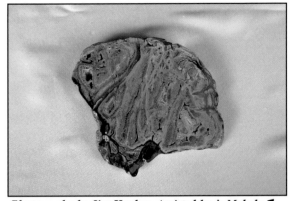

Photography by Jim Hughes, Assisted by ♪ Melody ♫
Collection of ♪ Melody ♫, Applewood, CO, USA

KORNERUPINE - $Mg_4Al_6(Si,Al,B)_5O_{21}(OH)$ - *Vitreous white, colourless, yellow; Hardness 6.5; Locality: Quebec, Canada*

Photography by Jim Hughes, Assisted by ♪ Melody ♫
Collection of ♪ Melody ♫, Applewood, CO, USA

KUNZITE - $LiAl(SiO_3)_2$ - *Vitreous pink to lilac colour; Hardness 6.5-7; Locality: Afghanistan*

Photography by Jim Hughes, Assisted by ♪ Melody ♫
Collection of ♪ Melody ♫, Applewood, CO, USA

KYANITE - Al_2SiO_5 - *Translucent blue, tan, white, grey, <u>green</u>, yellow, pink, black, etc.; Hardness 5.5-7; Locality: Minas Gerais, Brasil.*

Photography by Jim Hughes, Assisted by ♪ Melody ♫
Collection of Jackson & Melody, Applewood, CO, USA

KYANITE - Al_2SiO_5 - *Translucent blue, <u>tan</u>, white, grey, green, yellow, pink, black, etc.; Hardness 5.5-7; Locality: Minas Gerais, Brasil.*

Photography by Jim Hughes, Assisted by ♪ Melody ♫
Collection of ♪ Melody ♫, Applewood, CO, USA

KYANITE - Al_2SiO_5 - *Translucent blue, tan, white, grey, green, yellow, pink, <u>black</u>, etc.; Hardness 5.5-7; Locality: Minas Gerais, Brasil.*

Photography by Jim Hughes, Assisted by ♪ Melody ♫
Collection of ♪ Melody ♫, Applewood, CO, USA

KYANITE *- Al₂SiO₅ - (Faceted and Natural) - Translucent* <u>*blue*</u>*, tan, white, grey, green, yellow, pink, black, etc.; Hardness 5.5-7; Minas Gerais, Brasil.*
Photography by Jim Hughes, Assisted by ♪ Melody ♫
Collection of ♪ Melody ♫, Applewood, CO, USA

LABRADORITE -
*(Ca,Na)(Si,Al)$_4$O$_8$; Pearly
vitreous grey, brown, greenish,
colourless, etc.; Hardness 5-6;
Locality: Newfoundland,
Canada*

*Photography by Jim Hughes, Assisted by ♪ Melody ♫
Collection of Bob Jackson, Applewood, CO, USA*

LABRADORITE -
*(Ca,Na)(Si,Al)$_4$O$_8$; Pearly
vitreous grey, brown, greenish,
colourless, etc.; Hardness 5-6;
Locality: Newfoundland,
Canada*

*Photography by Jim Hughes, Assisted by ♪ Melody ♫
Collection of Bob Jackson, Applewood, CO, USA*

LABRADORITE -
*(Ca,Na)(Si,Al)$_4$O$_8$; Pearly
vitreous grey, brown, greenish,
colourless, etc.; Hardness 5-6;
Locality: Newfoundland,
Canada*

*Photography by Jim Hughes, Assisted by ♪ Melody ♫
Collection of Bob Jackson, Applewood, CO, USA*

LAMPROPHYLLITE -
$(Na,Ca)(Na,Mn)_2(Sr,Ba)_2Ti_3(Si_2O_7)_2(O,OH,F)_4$ - *Translucent golden-brown, brown; Hardness 2-3; Locality: Kola Peninsula, Russia*

Photography by Jim Hughes, Assisted by ♪ Melody ♫
Collection of ♪ Melody ♫, Applewood, CO, USA

LANTHANITE (On Cerite) -
$(La,Ce)_2(CO_3)_3$,♥$8H_2O$ - *Pearly colourless, white, pink, yellowish; Hardness 2.5-3; Locality: Roddarjuttam, Sweden*

Photography by Jim Hughes, Assisted by ♪ Melody ♫
Collection of ♪ Melody ♫, Applewood, CO, USA

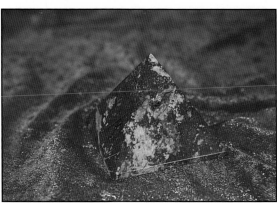

LAPIS LAZULI -
$3NaAlSiO_4$♥Na_2S *with Pyrite, Lazurite, Koksharovite, Muscovite, Calcite, and a Diopside free from Iron; Vitreous blue, violet-blue, greenish-blue; Hardness 5-5.5; Locality: Badakhshan, Afghanistan*

Photography by Jim Hughes, Assisted by ♪ Melody ♫
Collection of ♪ Melody ♫, Applewood, CO, USA

LAPIS LAZULI -
$3NaAlSiO_4 \heartsuit Na_2S$ with Pyrite, Lazurite, Koksharovite, Muscovite, Calcite, and a Diopside free from Iron; Vitreous blue, violet-blue, greenish-blue; Hardness 5-5.5; Locality: Badakhshan, Afghanistan

Photography by Jim Hughes, Assisted by ♪ Melody ♫
Collection of ♪ Melody ♫, Applewood, CO, USA

LARIMAR STONE -
$NaCa_2Si_3O_8(OH)$ with impurities (blue pectolite); Hardness 5; Locality: Dominican Republic

Photography by Jim Hughes, Assisted by ♪ Melody ♫
Collection of Bob Jackson, Applewood, CO, USA

LARSENITE - $ZnPbSiO_4$ -
Adamantine white to yellowish; Hardness 6.5-7; Locality: New Jersey, USA

Photography by Jim Hughes, Assisted by ♪ Melody ♫
Collection of ♪ Melody ♫, Applewood, CO, USA

LAUBMANNITE -
$Fe_9(PO_4)_4(OH)_{12}$ - *Vitreous silky*
<u>*greyish-green*</u>, *greenish brown,*
brown; Hardness 3.5-4; Locality:
Arkansas, USA

Photography by Jim Hughes, Assisted by ♪ Melody
Collection of ♪ Melody ♫, Applewood, CO, USA

LAUEITE - *(Crystals, With*
White/Yellow Hair-Like Strunzite
Crystals); $MnFe_2(PO_4)_2(OH)_2$♥
$8H_2O$ *-Translucent honey-brown*
wedges; Hardness 3; Locality:
New Hampshire, USA

Photography by Jim Hughes, Assisted by ♪ Melody ♫
Collection of ♪ Melody ♫, Applewood, CO, USA

LAUMONTITE -
$(Ca,Na_2)Al_2Si_4O_{12}$♥$4H_2O$; <u>*Vitreous*</u>
<u>*white*</u> *to yellow or grey, sometimes*
red; Hardness 3.5-4; Locality:
Bombay, India

Photography by Jim Hughes, Assisted by ♪ Melody ♫
Collection of ♪ Melody ♫, Applewood, CO, USA

LAVENITE -
Na(Mn,Ca,Fe)(Zr,Nb)(Si$_2$O$_7$)O,F)$_2$ - Vitreous light yellow; Hardness 6; Locality: Langesundfjord, Norway

Photography by Jim Hughes, Assisted by ♪ Melody ♫
Collection of ♪ Melody ♫, Applewood, CO, USA

LAZULITE -
(Mg,Fe)Al$_2$(PO$_4$)$_2$(OH)$_2$ - Vitreous blue, bluish-green; Hardness 5.5-6; Locality: Yukon, Canada

Photography by Jim Hughes, Assisted by ♪ Melody ♫
Collection of ♪ Melody ♫, Applewood, CO, USA

LAZULITE -
(Mg,Fe)Al$_2$(PO$_4$)$_2$(OH)$_2$ - Vitreous blue, bluish-green; Hardness 5.5-6; Locality: Yukon, Canada

Photography by Jim Hughes, Assisted by ♪ Melody ♫
Collection of Colorado School Of Mines, Colorado, USA

LAZURITE - $Na_3Ca(Si_3Al_3)O_{12}S$ -
*Vitreous blue, violet-blue,
greenish-blue; Hardness 5-5.5;
Locality: Badakhstan, Afghanistan*

*Photography by Jim Hughes, Assisted by ♪ Melody ♫
Collection of ♪ Melody ♫, Applewood, CO, USA*

LAZURITE - $Na_3Ca(Si_3Al_3)O_{12}S$ -
*Vitreous blue, violet-blue,
greenish-blue; Hardness 5-5.5;
Locality: Coquimbo, Chile, South
America*

*Photography by Jim Hughes, Assisted by ♪ Melody ♫
Collection of ♪ Melody ♫, Applewood, CO, USA*

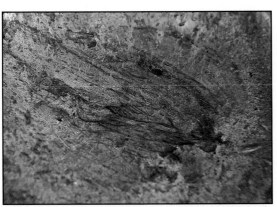

LEAD - Pb - *Metallic grey,
greyish-white in reflected light;
Hardness 1.5; Locality: Pushkara*

*Photography by Jim Hughes, Assisted by ♪ Melody ♫
Collection of Bob Jackson, Applewood, CO, USA*

LEAVERITE - *Multiple compositions, hardness, and colours; Locations throughout the world.*

Photography by Jim Hughes, Assisted by ♪ Melody ♫
Collection of ♪ Melody ♫, Applewood, CO, USA

LEGRANDITE -
$Zn_2AsO_4(OH)$ ♥ H_2O -
Translucent colourless, yellow; Hardness 5; Locality: Mexico

Photography by Jim Hughes, Assisted by ♪ Melody ♫
Collection of ♪ Melody ♫, Applewood, CO, USA
Gift of Angel Torricillas, Mexico

LEGRANDITE - *(On Limonite)*
$Zn_2AsO_4(OH)$ ♥ H_2O -
Translucent colourless, yellow; Hardness 5; Locality: Mexico

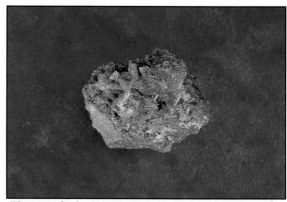

Photography by Jim Hughes, Assisted by ♪ Melody ♫
Collection of ♪ Melody ♫, Applewood, CO, USA

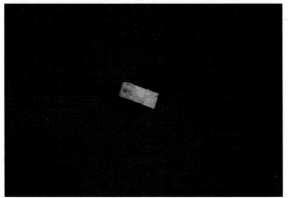

LEIFITE -
$Na_6Be_2Al_2Si_{16}O_{39}(OH)_2 \heartsuit 1.5H_2O$ -
Translucent colourless; Hardness of 6; Locality: Quebec, Canada

Photography by Jim Hughes, Assisted by ♪ Melody ♫
Collection of ♪ Melody ♫, Applewood, CO, USA

LEPIDOCROCITE - $Fe_2O_3 \heartsuit H_2O$
dimorphous - Adamantine yellows, reds, browns, etc.; Hardness 5-5.5; Locality: Spain

Photography by Jim Hughes, Assisted by ♪ Melody ♫
Collection of ♪ Melody ♫, Applewood, CO, USA

LEPIDOLITE -
$K(Li,Al)_3(Si,Al)_4O_{10}(F,OH)_2$ - *Pink, purple, colourless; Hardness from 2.5-4; Locality: California, USA*

Photography by Jim Hughes, Assisted by ♪ Melody ♫
Collection of ♪ Melody ♫, Applewood, CO, USA

LEUCITE (In Lava) -
$K(AlSi_2)O_6$ - Vitreous white, grey; Hardness 5.5-6; Locality: Campania, Italy

Photography by Jim Hughes, Assisted by ♪ Melody ♫
Collection of ♪ Melody ♫, Applewood, CO, USA

LEUCOPHOENICITE - (Pink, with Willemite and Zincite in Franklinite [black])-
$Mn_7(SiO_4)_3(OH)_2$ - Translucent light purplish-red; Hardness from 5.5-6; Locality: New Jersey, USA

Photography by Jim Hughes, Assisted by ♪ Melody ♫
Collection of ♪ Melody ♫, Applewood, CO, USA

LEVYNITE - $CaAl_2Si_3O_{10}♥5H_2O$ - Vitreous transparent to translucent colourless, white, greyish, reddish, yellowish, etc.; Hardness 4-4.5; Locality: Colorado, USA

Photography by Jim Hughes, Assisted by ♪ Melody ♫
Collection of Colorado School Of Mines, Colorado, USa

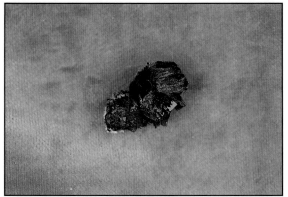

LIBETHENITE - $Cu_2PO_4(OH)$ - *Vitreous light/dark olive-green, deep green, blackish-green; Hardness 4; Locality, Zambia, Africa*

Photography by Jim Hughes, Assisted by ♪ Melody ♫
Collection of ♪ Melody ♫, Applewood, CO, USA

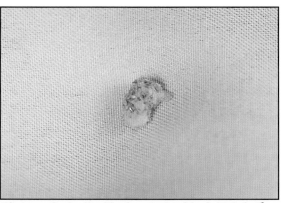

LIEBIGITE - $Ca_2(UO_2)(CO_3)_3$♥$11H_2O$ - *Vitreous green, yellowish-green; Hardness from 2.5-3; Locality: Colorado, USA*

Photography by Jim Hughes, Assisted by ♪ Melody ♫
Collection of ♪ Melody ♫, Applewood, CO, USA

LIMB CAST - *Fossilized tree limb where wood structure has been replace with agate and/or opal; Hardness variable.*

Photography by Jim Hughes, Assisted by ♪ Melody ♫
Collection of Bob Jackson, Applewood, CO, USA

LIMB CAST - *Fossilized tree limb where wood structure has been replace with agate and/or opal; Hardness variable.*

Photography by Jim Hughes, Assisted by ♪ Melody ♫
Collection of Bob Jackson, Applewood, CO, USA

LIMONITE - *$2Fe_2O_3$♥$3H_2O$ - Silky to sub-metallic or dull/earthy brown-yellow, ocher-yellow, with black exterior occasionally; Hardness 5-5.5; Localilty: Namibia, Africa*

Photography by Jim Hughes, Assisted by ♪ Melody ♫
Collection of ♪ Melody ♫, Applewood, CO, USA

LIMONITE - *(After Siderite); $2Fe_2O_3$♥$3H_2O$ - Silky to sub-metallic or dull/earthy brown-yellow, ocher-yellow, with black exterior occasionally; Hardness 5-5.5; Localilty: Colorado, USA*

Photography by Jim Hughes, Assisted by ♪ Melody ♫
Collection of ♪ Melody ♫, Applewood, CO, USA

LINARITE (On Fluorite) - $CuPbSO_4(OH)_2$ **- Vitreous/sub-adamantine blue; Hardness 2.5; Locality: Jaen, Spain**

Photography by Jim Hughes, Assisted by ♪ Melody ♫
Collection of ♪ Melody ♫, Applewood, CO, USA

LINNAEITE - Co_3S_4 **- Metallic grey; Hardness 4.5-5.5; Locality: Missouri, USA**

Photography by Jim Hughes, Assisted by ♪ Melody ♫
Collection of Colorado School Of Mines, Colorado, USA

LODESTONE - Fe_3O_4 **with polar magnetism - Metallic to sub-metallic iron-black; Hardness from 5.5 - 6.5; Arkansas, USA**

Photography by Jim Hughes, Assisted by ♪ Melody ♫
Collection of ♪ Melody ♫, Applewood, CO, USA

LOMONOSOVITE -
$Na_5Ti_2O_2(Si_2O_7)(PO_4)$ -
*Vitreous/adamantine dark
brown, black, rose-violet, etc.;
Hardness 3-4; Locality: Kola
Peninsula, Russia*

Photography by Jim Hughes, Assisted by ♪ Melody ♫
Collection of ♪ Melody ♫, Applewood, CO, USA

LUDLAMITE -
$(Fe,Mg,Mn)_3(PO_4)_2 \heartsuit 4H_2O$ -
*Vitreous green; Hardness 3.5;
Locality: Arizona, USA*

Photography by Jim Hughes, Assisted by ♪ Melody ♫
Collection of ♪ Melody ♫, Applewood, CO, USA

LUDLAMITE -
$(Fe,Mg,Mn)_3(PO_4)_2 \heartsuit 4H_2O$ -
*Vitreous green; Hardness 3.5;
Locality: Cornwall, England*

Photography by Jim Hughes, Assisted by ♪ Melody ♫
Collection of Bob Jackson, Applewood, CO, USA

MAGNESIOFERRITE - *(Black Crystals); MgFe₂O₄ - Metallic/ sub-metallic black, grey; Hardness 5.5-6.5; Locality: Sweden*

Photography by Jim Hughes, Assisted by ♩ Melody ♫ Collection of ♩ Melody ♫, Applewood, CO, USA

MAGNESITE - *MgCO₃ - Vitreous colourless, <u>white,</u> greyish-white, yellowish, brown; Hardness 4; Locality: Arizona, USA*

Photography by Jim Hughes, Assisted by ♩ Melody ♫ Collection of Bob Jackson, Applewood, CO, USA

MAGNESITE - *MgCo₃; Vitreous white, grey, yellow, brown; Hardness 3.5-4.5; Locality; Bahia, Brasil*

Photography by Jim Hughes, Assisted by ♩ Melody ♫ Collection of ♩ Melody ♫, Applewood, CO, USA

MAGNETITE - Fe_3O_4 - *Metallic black, brownish-grey; Hardness 5.5-6.5; Locality: Minas Gerais, Brasil*

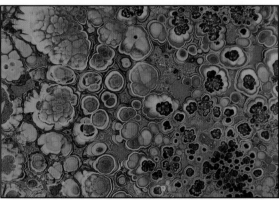

MALACHITE - $Cu_2CO_3(OH)_2$ - *Adamantine/vitreous/silky/earthy green; Hardness 3.5-4; Locality: Zaire*

MANGANESE (*Ore*) - *Mn; Vitreous grey-black to black mass; Hardness: 4: Locality: Republic of South Africa, Africa*

MANGANITE - (Crystals)- alpha-MnO(OH) - Sub-metallic grey, black; Hardness 4; Ontario, Canada

Photography by Jim Hughes, Assisted by ♪ Melody ♫
Collection of ♪ Melody ♫, Applewood, CO, USA

MANGANOSITE - (Coating); MnO; Vitreous green; Hardness 5.5; Locality: Sweden

Photography by Jim Hughes, Assisted by ♪ Melody ♫
Collection of Bob Jackson, Applewood, CO, USA

MANGANOSITE - (Green with Zincite); MnO; Vitreous green; Hardness 5.5; Locality: New Jersey, USA

Photography by Jim Hughes, Assisted by ♪ Melody ♫
Collection of Colorado School Of Mines, CO, USA

MARBLE - $CaCo_3$ with impurities; White shiny; Metamorphosed limestone of coarse to medium-grained rock of re-crystallized carbonates; Hardness variable; Locality: India (Same material used to build the Taj Mahal)

Photography by Dave Shrum, Colorado Camera Co., Lakewood, Colorado, USA
Collection of Bob Jackson, Applewood, CO, USA

MARCASITE - FeS_2 - Metallic white/bronze-yellow; Hardness from 6-6.5; Locality: France

Photography by Jim Hughes, Assisted by ♪ Melody ♫
Collection of Bob Jackson, Applewood, CO, USA

MARCASITE - (Nodule); FeS_2 ; Metallic white/bronze-yellow; Hardness from 6-6.5; Locality: France

Photography by Jim Hughes, Assisted by ♪ Melody ♫
Collection of Bob Jackson, Applewood, CO, USA

MARCASITE - *(In Agate); FeS₂ Metallic white/bronze-yellow; Hardness from 6-6.5; Locality: California, USA*

Photography by Jim Hughes, Assisted by ♪ Melody ♫
Collection of ♪ Melody ♫, Applewood, CO, USA

MARGARITE - *(With Corundum) - $CaAl_2(Si_2Al_2)O_{10}(OH)_2$ - Translucent greyish-pink, pale yellow/green; Hardness 3.5-4.5; Locality: Tyrol, Austria*

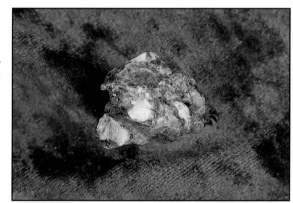

Photography by Jim Hughes, Assisted by ♪ Melody ♫
Collection of ♪ Melody ♫, Applewood, CO, USA

MARIALITE - *$Na_4(Si_3Al)_{12}O_{24}Cl$ - Vitreous colourless, white; Hardness 5.5-6; Locality: Naples, Italy*

Photography by Jim Hughes, Assisted by ♪ Melody ♫
Collection of ♪ Melody ♫, Applewood, CO, USA

MATLOCKITE - *(Glassy yellow crystals) - PbClF - Adamantine colourless, <u>yellow</u>, pale amber, greenish; Hardness 2.5-3; Derbyshire, England*

Photography by Jim Hughes, Assisted by ♪ Melody Collection of ♪ Melody ♫, Applewood, CO, USA

MELANITE - $3CaO♥Fe_2O_3♥3SiO_2$; *Dull or lustrous black calcium-iron garnet; Hardness 6.5 - 7.5; Locality: California, USA*

Photography by Jim Hughes, Assisted by ♪ Melody ♫ Collection of ♪ Melody ♫, Applewood, CO, USA

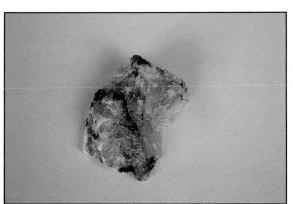

MELIPHANE - $(Ca,Na)_2Be(Si,Al)_2(O,F)_7$ - *Yellow to colourless; Hardness 5-5.5; Locality: Langesundsfjord, Norway*

Photography by Jim Hughes, Assisted by ♪ Melody ♫ Collection of ♪ Melody ♫, Applewood, CO, USA

MESOLITE -
$Na_2Ca_2(Al_6Si_9)O_{30} \heartsuit 8H_2O$ -
Vitreous/silky white, colourless,
greyish, yellowish; Hardness 5;
Locality: Pune, India

Photography by Jim Hughes, Assisted by ♪ Melody ♫
Collection of ♪ Melody ♫, Applewood, CO, USA

MESSELITE -
$Ca_2(Fe,Mn)(PO_4)_2 \heartsuit 2H_2O$ -
Vitreous greenish white/grey;
Hardness 3.5; Locality: Hessen,
Germany

Photography by Jim Hughes, Assisted by ♪ Melody ♫
Collection of ♪ Melody ♫, Applewood, CO, USA

METEORITE - *Composition,*
colour range, and hardness
variable; Locality: Texas, USA

Photography by Jim Hughes, Assisted by ♪ Melody ♫
Collection of Bob Jackson, Applewood, CO, USA

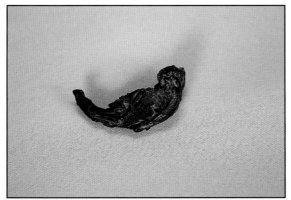

METEORITE - *(Irghisite - Impact Glass); Composition, colour range, and hardness variable; Locality: Z-Center, Russia*

Photography by Jim Hughes, Assisted by ♪ Melody ♫
Collection of ♪ Melody ♫, Applewood, CO, USA

MICA - *(Leaf) - (K,Na,Ca, Ba,H_3O,NH_4)(Al,Mg,Fe,Li,Cr, Mn,V,Zn)$_{2-3}$(Si,Al,Fe)$_4O_{10}$(OH,F)$_2$ - [General sheet silicate]; Hardness 2-2.5; Locality: Washington, USA*

Photography by Jim Hughes, Assisted by ♪ Melody ♫
Collection of Bob Jackson, Applewood, CO, USA

MICROLITE - *(In Albite) - (Ca,Na)$_2$Ta$_2O_6$(O,OH,F) - Vitreous/resinous <u>yellow/green</u>, brown; Hardness 5-5.5; Locality: Connecticut, USA*

Photography by Jim Hughes, Assisted by ♪ Melody ♫
Collection of ♪ Melody ♫, Applewood, CO, USA

MILARITE -
$(K,Na)Ca_2(Be,Al)_3Si_{12}O_{30} \heartsuit H_2O$ -
Vitreous colourless, <u>white</u>, pale green; Hardness 5.5-6; Locality: Minas Gerais, Brasil

Photography by Jim Hughes, Assisted by ♪ Melody ♫
Collection of ♪ Melody ♫, Applewood, CO, USA

MILLERITE - beta-NiS -
Metallic; Hardness 3-3.5; Locality: Wisconsin, USA

Photography by Jim Hughes, Assisted by ♪ Melody ♫
Collection of ♪ Melody ♫, Applewood, CO, USA

MIMETITE - $Pb_5(AsO_4)_3Cl$ -
Resinous/sub-adamantine <u>yellow</u>, orange, white, etc.; Hardness 3.5-4; Locality: Namibia, Africa

Photography by Jim Hughes, Assisted by ♪ Melody ♫
Collection of ♪ Melody ♫, Applewood, CO, USA

"MINERAL STONE" - *Contains over 17 different minerals; Hardness variable; Locality: North Carolina, USA*

Photography by Jim Hughes, Assisted by ♪ Melody ♫
Collection of ♪ Melody ♫, Applewood, CO, USA

MITRIDATITE - $Ca_2Fe_3O_2(PO_4)_3$♥$3H_2O$ - *Earthy/resinous greenish-yellow, green, brownish-green, etc.; Hardness: 2.5; Locality: New Hampshire, USA*

Photography by Jim Hughes, Assisted by ♪ Melody ♫
Collection of Colorado School Of Mines, Colorado, USA

MIXITE - $Cu_6Bi(AsO_4)_3(OH)_6$♥$3H_2O$ - *Dull to brilliant green, bluish-green, pale green, whitish; Hardness 3-4; Locality: Utah, USA*

Photography by Jim Hughes, Assisted by ♪ Melody ♫
Collection of ♪ Melody ♫, Applewood, CO, USA

MOHAWKITE - Cu_3As with Ni and Co - Metallic <u>lustrous</u> grey to <u>yellow-brown</u>; Hardness 3.5; Locality: Michigan, USA

Photography by Jim Hughes, Assisted by ♪ Melody ♫
Collection of Bob Jackson, Applewood, CO, USA

MOLDAVITE - Tektite (green and glassy) from River Valley in Czechoslovakia; Known also as Vltava; Hardness estimated at approximately 5.

Photograph courtesy of Bob Simmons and Kathy Warner; Photography by David Benoit, Massachusetts, USA

MONAZITE - Crystals) - $(Nd,La,Ce)PO_4$ - Translucent to milky rose to <u>red-brown</u>, grey/ blue; Hardness 5-5.5; Locality: Maine, USA

Photography by Jim Hughes, Assisted by ♪ Melody ♫
Collection of ♪ Melody ♫, Applewood, CO, USA

MONAZITE - Crystals) - (Nd,La,Ce)PO_4 - Translucent to milky rose to red-brown, <u>grey/ blue</u>; Hardness 5-5.5; Locality: Japan

Photography by Jim Hughes, Assisted by ♪ Melody ♫
Collection of ♪ Melody ♫, Applewood, CO, USA

MONTICELLITE - CaMgSiO_4 - Translucent colourless, grey; Hardness 5.5; Locality: California, USA

Photography by Jim Hughes, Assisted by ♪ Melody ♫
Collection of ♪ Melody ♫, Applewood, CO, USA

MOONSTONE - Combination of Orthoclase and Albite - Chatoyant opalescent reflection or schiller; Hardness 6; Locality: Sri Lanka

Photography by Jim Hughes, Assisted by ♪ Melody ♫
Collection of ♪ Melody ♫, Applewood, CO, USA

MORDENITE - *(Needles, with blue-green Caledonite)* - $K_{2.8}Na_{1.5}Ca_2(Al_9Si_{39})O_{96} \cdot 29H_2O$ - *Vitreous/pearly/silky white, yellowish, pinkish; Hardness from 3-4; Locality: Nova Scotia, Canada*

*Photography by Jim Hughes, Assisted by ♪ Melody ♫
Collection of ♪ Melody ♫, Applewood, CO, USA*

MORGANITE - $Be_3Al_2Si_6O_{18}$ - *Vitreous rose to peach colour; Hardness 7.5-8; Locality: Minas Gerais, Brasil*

*Photography by Jim Hughes, Assisted by ♪ Melody ♫
Collection of ♪ Melody ♫, Applewood, CO, USA*

MOSANDRITE - $(Ca,Na,Ce)_{12}(Ti,Zr)_2Si_7O_{25}(OH)_6F_4$ - *Resinous/vitreous lustrous reddish-brown, greenish, etc.; Hardness 4; Locality: Quebec, Canada*

*Photography by Jim Hughes, Assisted by ♪ Melody ♫
Collection of ♪ Melody ♫, Applewood, CO, USA*

MOTTRAMITE - *(On Calcite) - Resinous luster, olive to deep green, translucent to sub-translucent; Hardness 3.5; A type of Descloizite where the zinc is almost entirely replaced by copper; Locality: Namibia, Africa*

Photography by Jim Hughes, Assisted by ♪ Melody ♫
Collection of ♪ Melody ♫, Applewood, CO, USA

MUIRITE - $Ba_{10}Ca_2MnTiSi_{10}O_{30}(OH,Cl,F)_{10}$ - *Sub-vitreous orange to yellow-orange; Hardness 2.5; Locality: California, USA*

Photograph by alkdjf;laksjf
Collection of a;ldfkj;aljf

MULLITE - $3Al_2O_3 \heartsuit 2SiO_2$ - *Vitreous pale pink; Hardness 6-7; Locality: Island of Mull, Scotland*

Photography by Jim Hughes, Assisted by ♪ Melody ♫
Collection of ♪ Melody ♫, Applewood, CO, USA

MUSCOVITE - (Green);
(H,K)AlSiO₄ ; Hardness 2 - 2.5;
Locality: Minas Gerais, Brasil

Photography by Jim Hughes, Assisted by ♪ Melody ♫
Collection of ♪ Melody ♫, Applewood, CO, USA

MUSCOVITE - (Rose);
(H,K)AlSiO₄ ; Hardness 2 - 2.5;
Locality: New Mexico, USA

Photography by Jim Hughes, Assisted by ♪ Melody ♫
Collection of ♪ Melody ♫, Applewood, CO, USA

MUSCOVITE - (On Quartz);
(H,K)AlSiO₄ ; Hardness 2 - 2.5;
Locality: South Dakota, USA

Photography by Jim Hughes, Assisted by ♪ Melody ♫
Collection of ♪ Melody ♫, Applewood, CO, USA

NADORITE - *(Crystals)* - *$PbSbO_2Cl$* - *Resinous/ adamantine brown, yellow; Hardness 3.5-4; Locality: Constantine, Algeria*

Photography by Jim Hughes, Assisted by ♪ Melody ♫
Collection of ♪ Melody ♫, Applewood, CO, USA

NARSARSUKITE - *$Na_2(Ti,Fe)Si_4(O,F)_{11}$* - *Translucent <u>yellow</u>, brownish-grey; Hardness 7; Locality, Quebec, Canada*

Photography by Jim Hughes, Assisted by ♪ Melody ♫
Collection of ♪ Melody ♫, Applewood, CO, USA

NATROLITE - *$Na_2(Al_2Si_3)O_{10}$♥$2H_2O$* - *Vitreous/<u>pearly white</u>, colourless, greyish, yellowish, etc.; Hardness 5-5.5; Locality: California, USA*

Photography by Jim Hughes, Assisted by ♪ Melody ♫
Collection of ♪ Melody ♫, Applewood, CO, USA

NEPHELINE *(Soda) - (Na,K)AlSiO₄ (Also Known As Belocilite) - Vitreous/lustrous white, yellowish, green, grey, etc.; Hardness 3-5; Locality: Quebec, Canada*

Photography by Jim Hughes, Assisted by ☽ Melody ♫
Collection of ☽ Melody ♫, Applewood, CO, USA

NEPHELITE - *(Green, with Vesuvianite, Cuspidine) - Vitreous luster, colourless, white, yellowish, green, blue-grey, brown-red, red; Hardness 5.5-6; Locality: Vesuvius, Italy.*

Photography by Jim Hughes, Assisted by ☽ Melody ♫
Collection of Colorado School Of Mines, CO, USA

NEPHRITE - *Ca₂(Mg,Fe)₅(OH)₂(Si₄O₁₁)₂ - (Variety: Alaskan Beach Jade) - Compact fine-grained mineral coloured <u>green</u>, cream, grey, pink, etc.; Hardness 6-6.5; Locality: Alaska, USA*

Photography by Jim Hughes, Assisted by ☽ Melody ♫
Collection of Bob Jackson, Applewood, CO, USA

NEPHRITE -
$Ca_2(Mg,Fe)_5(OH)_2(Si_4O_{11})_2$ -
(Variety: Black) - Compact fine-grained mineral; Hardness 6-6.5; Locality: Wyoming, USA

Photography by Jim Hughes, Assisted by ♪ Melody ♫
Collection of Bob Jackson, Applewood, CO, USA

NEPHRITE -
$Ca_2(Mg,Fe)_5(OH)_2(Si_4O_{11})_2$ -
(Variety: Monterey Jade) - Compact fine-grained mineral coloured <u>green</u>, cream, grey, pink, etc.; Hardness 6-6.5; Locality: California, USA

Photography by Jim Hughes, Assisted by ♪ Melody ♫
Collection of Bob Jackson, Applewood, CO, USA

NEPHRITE -
$Ca_2(Mg,Fe)_5(OH)_2(Si_4O_{11})_2$ -
(Variety: Columbia River Jade) - Compact fine-grained mineral coloured <u>green to black</u>, cream, grey, pink, etc.; Hardness 6-6.5; Locality: Washington, USA

Photography by Jim Hughes, Assisted by ♪ Melody ♫
Collection of Bob Jackson, Applewood, CO, USA

NEPHRITE QUARTZ - $Ca_2(Mg,Fe)_5(OH)_2(Si_4O_{11})_2$ *With* SiO_2 - *Compact fine-grained green, with quartz crystal structures; Hardness 6-7; Locality: Wyoming, USA Photography by Jim Hughes, Assisted by ♪ Melody ♫; Collection of ♪ Melody ♫*

NEPTUNITE (Crystals) - $KNa_2Li(Fe,Mg,Mn)_2Ti_2Si_8O_{24}$ - *Black; Hardness 5-6; California, USA. Photography by Jim Hughes, Assisted by ♪ Melody ♫; Collection of ♪ Melody ♫*

NICCOLITE (With Annabergite) - *NiAs; Metallic lustrous pale copper-red; Hardness 5-5.5; Locality: Nevada, USA*

Photography by Jim Hughes, Assisted by ♪ Melody ♫
Collection of ♪ Melody ♫, Applewood, CO, USA

NISSONITE - *$CuMgPO_4(OH)$♥$2.5H_2O$ - Translucent bluish-green; Locality: California, USA*

Photography by Jim Hughes, Assisted by ♪ Melody ♫
Collection of ♪ Melody ♫, Applewood, CO, USA

NORBERGITE - *$Mg_3(SiO_4)(F,OH)_2$ - Translucent tawny; Hardness 6.5; Locality: New Jersey, USA*

Photography by Jim Hughes, Assisted by ♪ Melody ♫
Collection of ♪ Melody ♫, Applewood, CO, USA

NORDENSKIOLDINE - *CaSn(BO$_3$)$_2$ - Vitreous colourless, yellow; Hardness 5.5-6; Locality: Yukon, Canada*

Photography by Jim Hughes, Assisted by ♪ Melody ♫
Collection of ♪ Melody ♫, Applewood, CO, USA

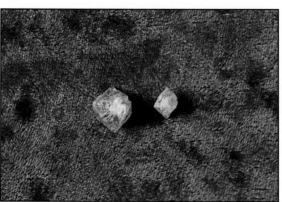

NORTHUPITE - *Na$_3$Mg(CO$_3$)$_2$Cl - Vitreous colourless, pale yellow, grey, brown; Hardness 3.5-4; Locality: California, USA*

Photography by Jim Hughes, Assisted by ♪ Melody ♫
Collection of ♪ Melody ♫, Applewood, CO, USA

NUUMMIT - *Combination of (Mg,Fe)$_7$(Si$_8$O$_{22}$)(OH,F)$_2$ [Anthophyllite] and (Mg,Fe)$_5$(Al$_2$O$_{22}$)(OH,F$_2$) [Gedrite] - Laminate crystallization, opaque black with iridescent spectral colour range; Hardness 5.5-6; Locality: Southern Greenland*

Photography by Jim Hughes, Assisted by ♪ Melody ♫
Collection of Bob Jackson, Applewood, CO, USA

*Do not fear.
But if you must fear,
fear the "known" rather than the unknown,
because the "known" limits the perception
(or acceptance) of the unknown.*

W. Lynn McKinney, PhD, Cocoa Beach, FL, USA

OBSIDIAN - *(Apache Tear)*
Volcanic bright and vitreous
lustrous glass; Hardness 5-5.5.
Colour due to dustlike particles
of Magnetite and/or oxidized
Magnetite/Hematite; spherical
shape with conchoidal markings.
Locality: Arizona, USA

Photography by Jim Hughes, Assisted by ♪ Melody ♫
Collection of ♪ Melody ♫, Applewood, CO, USA

OBSIDIAN - *(Black)* *Volcanic*
bright and vitreous lustrous
glass; Hardness 5-5.5; Colour
due to dustlike particles of
Magnetite; Locality; Oregon,
USA

Photography by Jim Hughes, Assisted by ♪ Melody ♫
Collection of Bob Jackson, Applewood, CO, USA

OBSIDIAN - *(Black, Brown)*
Volcanic bright and vitreous
lustrous glass; Hardness 5-5.5;
Colour due to dustlike particles
of Magnetite and/or oxidized
Magnetite or Hematite; Locality:
Oregon, USA

Photography by Jim Hughes, Assisted by ♪ Melody ♫
Collection of ♪ Melody ♫, Applewood, CO, USA

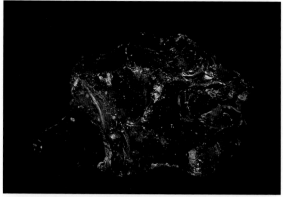

OBSIDIAN - (Blue-Green-Gold)
Volcanic bright and vitreous lustrous glass; Hardness 5-5.5. Colour due to inclusions not yet defined; Locality: Middle East

Photography by Jim Hughes, Assisted by ♪ Melody ♫
Collection of ♪ Melody ♫, Applewood, CO, USA

OBSIDIAN - (Blue) *Volcanic bright and vitreous lustrous glass; Hardness 5-5.5; Colour due to inclusions not yet defined; Locality: Middle East*

Photography by Jim Hughes, Assisted by ♪ Melody ♫
Collection of ♪ Melody ♫, Applewood, CO, USA

OBSIDIAN - (Electric Blue Sheen) (Cabachon); *Volcanic bright and vitreous lustrous glass; Hardness 5-5.5; Colour due to inclusions not yet defined; Locality: Oregon, USA*

Photography by Jim Hughes, Assisted by ♪ Melody ♫
Collection of ♪ Melody ♫, Applewood, CO, USA

OBSIDIAN - (Electric Blue Sheen) *(Sphere); Volcanic bright and vitreous lustrous glass; Hardness 5-5.5; Colour due to inclusions not yet defined; Locality: Oregon, USA. Photography by Jim Hughes, Assisted by ♪ Melody ♫. Collection of Jackson & ♪ Melody ♫*

OBSIDIAN - (Gold Sheen)
Volcanic bright and vitreous
lustrous glass; Hardness 5-5.5.
Colour due to inclusions not yet
defined; Locality: Oregon, USA

Photography by Jim Hughes, Assisted by ♪ Melody ♫
Collection of ♪ Melody ♫, Applewood, CO, USA

OBSIDIAN - (Green) Volcanic
bright and vitreous lustrous glass;
Hardness 5-5.5; Colour due to
inclusions not yet defined;
Locality: Middle East

Photography by Jim Hughes, Assisted by ♪ Melody ♫
Collection of Julianne Guilbault, Lakewood, CO, USA

OBSIDIAN - (Grey) Volcanic
bright and vitreous lustrous glass;
Hardness 5-5.5; Colour due to
oxidized Magnetite or Hematite:
Locality: Oregon, USA Locality:
Oregon, USA

Photography by Jim Hughes, Assisted by ♪ Melody ♫
Collection of ♪ Melody ♫, Applewood, CO, USA

OBSIDIAN - (Mahogany)
Volcanic bright and vitreous lustrous glass; Hardness 5-5.5. Mahogany colour due to oxidized Magnetite or Hematite; Locality, Mexico

Photography by Jim Hughes, Assisted by ♪ Melody ♫
Collection of ♪ Melody ♫, Applewood, CO, USA

OBSIDIAN - (Purple Sheen)
Volcanic bright and vitreous lustrous glass; Hardness 5-5.5. Colour due to inclusions not yet defined; Locality: Oregon, USA

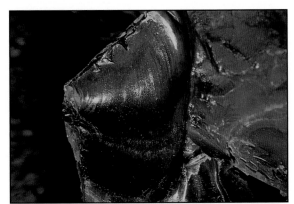

Photography by Jim Hughes, Assisted by ♪ Melody ♫
Collection of Bob Jackson, Applewood, CO, USA

OBSIDIAN - (Rainbow Sheen)
Volcanic bright and vitreous lustrous glass; Hardness 5-5.5. Colour due to inclusions oxidized Magnetite and/or Hematite and inclusions not yet defined; Locality: Oregon, USA

Photography by Jim Hughes, Assisted by ♪ Melody ♫
Collection of ♪ Melody ♫, Applewood, CO, USA

OBSIDIAN - (Red, Black, Snowflake) *Volcanic bright and vitreous lustrous glass; Hardness 5-5.5. See separate titles; Locality: Oregon/California, USA*

Photography by Jim Hughes, Assisted by ♪ Melody ♫
Collection of ♪ Melody ♫, Applewood, CO, USA
Gift from Bob Jackson, Colorado, USA

OBSIDIAN - (Silver Sheen) *Volcanic bright and vitreous lustrous glass; Hardness 5-5.5. Colour due to inclusions not yet defined; Locality: Mexico*

Photography by Jim Hughes, Assisted by ♪ Melody ♫
Collection of ♪ Melody ♫, Applewood, CO, USA

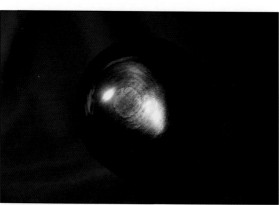

OBSIDIAN - (Snowflake) *Volcanic bright and vitreous lustrous glass; Hardness 5-5.5. Colour black due to oxidized Magnetite/Hematite; Snowflakes due to phenocrysts of possibly feldspar; Locality: California, USA*

Photography by Jim Hughes, Assisted by ♪ Melody ♫
Collection of ♪ Melody ♫, Applewood, CO, USA

OCHO - *(Type of very small Geode) - Mexico*

Photography by Jim Hughes, Assisted by ♪ Melody ♫
Collection of ♪ Melody ♫, Applewood, CO, USA

OKENITE - $Ca_{10}Si_{18}O_{46}♥18H_2O$ - *Sub/pearly* <u>white</u>, *yellowish, bluish; Hardness 4.5-5; Locality: Pune, India.*

Photography by Jim Hughes, Assisted by ♪ Melody ♫
Collection of ♪ Melody ♫, Applewood, CO, USA

OLIGOCLASE - *(With Twisted Quartz) - $(Na,Ca)(Si,Al)_4O_8$ - Vitreous/pearly whitish to pink and tan, greyish; Hardness 6-7; Locality: Minas Gerais, Brasil*

Photography by Jim Hughes, Assisted by ♪ Melody ♫
Collection of ♪ Melody ♫, Applewood, CO, USA

OLIVENITE - $Cu_2(OH)AsO_4$
Colour, various shades of olive green and brown, yellow, white; Sub-transparent to opaque; Hardness 3: Locality: Nevada, USA

Photography by Jim Hughes, Assisted by ♪ Melody ♫
Collection of ♪ Melody ♫, Applewood, CO, USA

ONYX - SiO_2 with impurities -
Even planed banded straight transparent to translucent white, grey, blue, brown, etc.; Hardness 7; Locality: Republic of South Africa, Africa

Photography by Jim Hughes, Assisted by ♪ Melody ♫
Collection of ♪ Melody ♫, Applewood, CO, USA

OPAL - (Andean) -SiO_2♥nH_2O
Hardness 5.5-6.5; Locality: Andes Mts., South America

Photography by Jim Hughes, Assisted by ♪ Melody ♫
Collection of ♪ Melody ♫, Applewood, CO, USA

OPAL - (Boulder Fire) - $SiO_2 \cdot nH_2O$ - Occurs within rounded or ellipsoidal-shaped boulders, like nodules or concretions which consist of silicious ironstone, with silica entering crevices; Hardness 5.5-6.5; Locality: Queensland, Australia

*Photography by ♪ Melody ♫, Colorado, USA
Collection of Bruce & Barbara McDougall, The Gold Coast, Queensland, Australia*

OPAL - (Boulder Fire) - $SiO_2 \cdot nH_2O$ - Occurs within rounded or ellipsoidal-shaped boulders, like nodules or concretions which consist of silicious ironstone, with silica entering crevices; Hardness 5.5-6.5; Locality: Queensland, Australia

*Photography by Jim Hughes, Assisted by ♪ Melody ♫
Collection of Bob Jackson, Applewood, CO, USA*

OPAL - (Boulder Matrix Fire) - $SiO_2 \cdot nH_2O$ - Opalized ironstone with random intermixing of gem opal with the brown ironstone; Hardness 5.5-6.5 (Opal); Locality: Queensland, Australia

*Photography by Jim Hughes, Assisted by ♪ Melody ♫
Collection of ♪ Melody ♫, Applewood, CO, USA
Gift of Bruce McDougall, Queensland, Australia*

OPAL - (Boulder Matrix Fire) - $SiO_2 \heartsuit nH_2O$ - Opalized ironstone with random intermixing of gem opal with the brown ironstone; Hardness 5.5-6.5 (Opal); Locality: Queensland, Australia

Photography by Jim Hughes, Assisted by ♪ Melody ♫
Collection of ♪ Melody ♫, Applewood, CO, USA
Gift of Bruce McDougall, Queensland, Australia

OPAL - (Boulder Matrix Fire) - $SiO_2 \heartsuit nH_2O$ - Opalized ironstone with random intermixing of gem opal with the brown ironstone; Hardness 5.5-6.5 (Opal); Locality: Queensland, Australia

Photography by Len Cram, Lightning Ridge,
New South Wales, Australia

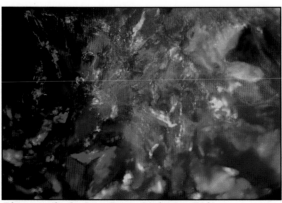

OPAL - (Black Opal - Lightning Ridge Fire) -$SiO_2 \heartsuit nH_2O$ - Formed as seam and small silicious nodules; Hardness 5.5-6.5; Locality: New South Wales, Australia

Photography by Len Cram, Lightning Ridge,
New South Wales, Australia

OPAL - (Black Opal - Lightning Ridge Fire) -SiO$_2$♥nH$_2$O - Formed as seam and small silicious nodules; Hardness 5.5-6.5; Locality: New South Wales, Australia

Photography by Len Cram, Lightning Ridge, New South Wales, Australia

OPAL - (Black Plume Opal) - SiO$_2$♥nH$_2$O - Formed as seam and small silicious nodules; Hardness 5.5-6.5; Locality: British Columbia, Canada

Photography by Jim Hughes, Assisted by ♪ Melody ♫ Collection of Bob Jackson, Applewood, CO, USA

OPAL - (Blue) -SiO$_2$♥nH$_2$O - Hardness 5.5-6.5; Locality: Peru, South America

Photography by Jim Hughes, Assisted by ♪ Melody ♫ Collection of ♪ Melody ♫, Applewood, CO, USA Gift from the "Mining Center", Peru, South America

OPAL - (Blue) -SiO$_2$♥nH$_2$O - Hardness 5.5-6.5; Locality: Mexico

Photography by Jim Hughes, Assisted by ♪ Melody ♫
Collection of Bob Jackson, Applewood, CO, USA

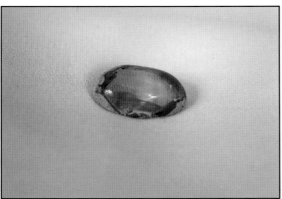

OPAL - (Cherry) -SiO$_2$♥nH$_2$O - Hardness 5.5-6.5; Locality: Mexico

Photography by Jim Hughes, Assisted by ♪ Melody ♫
Collection of ♪ Melody ♫, Applewood, CO, USA

OPAL - (Dendritic) -SiO$_2$♥nH$_2$O - Hardness 5.5-6.5; Locality: Georgia, USA

Photography by Jim Hughes, Assisted by ♪ Melody ♫
Collection of ♪ Melody ♫, Applewood, CO, USA

OPAL - (Fire - Jelly) -
SiO₂♥nH₂O - Hardness 5.5-6.5;
Locality: Oregon, USA

Photography by Jim Hughes, Assisted by ♩ Melody ♬
Collection of ♩ Melody ♬, Applewood, CO, USA

OPAL - (Fire - Golden) -
SiO₂♥nH₂O - Hardness 5.5-6.5;
Locality: Oregon, USA

Photography by Jim Hughes, Assisted by ♩ Melody ♬
Collection of Bob Jackson, Applewood, CO, USA

OPAL - (Fire) -SiO₂♥nH₂O -
Hardness 5.5-6.5; Locality:
Queensland, Australia

Photography by Jim Hughes, Assisted by ♩ Melody ♬
Collection of Bob Jackson, Applewood, CO, USA

OPAL - (Green) -SiO$_2$♥nH$_2$O - Hardness 5.5-6.5; Locality: Idaho, USA

Photography by Jim Hughes, Assisted by ♪ Melody ♫
Collection of ♪ Melody ♫, Applewood, CO, USA

OPAL - (Honduran) -SiO$_2$♥nH$_2$O - Hardness 5.5-6.5; Locality: Honduras

Photography by Jim Hughes, Assisted by ♪ Melody ♫
Collection of Bob Jackson, Applewood, CO, USA

OPAL - (Pink) -SiO$_2$♥nH$_2$O - Hardness 5.5-6.5; Locality: Idaho, USA

Photography by Jim Hughes, Assisted by ♪ Melody ♫
Collection of Bob Jackson, Applewood, CO, USA

OPAL - (Potch [White])-
$SiO_2 \heartsuit nH_2O$ - Hardness 5.5-6.5;
Locality: Idaho, USA

OPAL - (Red) -$SiO_2 \heartsuit nH_2O$ -
Hardness 5.5-6.5; Locality:
Oregon, USA

OPAL - (Water [Clear]) -
$SiO_2 \heartsuit nH_2O$ - Hardness 5.5-6.5;
Locality: Oregon, USA

***Opalized Fossilized Clam -
Western Australia***

***Opalized Wood - Washington,
USA***

***ORPIMENT - As_2S_3 ; Pearly to
resinous lemon to gold yellow;
sub- transparent to sub-
translucent; Hardness 1.5-2;
Locality: Nevada, USA***

ORPIMENT - As_2S_3 ; *Pearly to resinous lemon to gold yellow; sub- transparent to sub-translucent; Hardness 1.5-2; Locality: Utah, USA*

Photography by Jim Hughes, Assisted by ♪ Melody ♫
Collection of Bob Jackson, Applewood, CO, USA

ORTHOCLASE - $KAlSi_3O_8$ - *Vitreous colourless, white, pale yellow, pink, grey, etc., Hardness 6-6.5.; Locality: Madagascar*

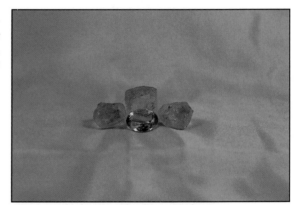

Photography by Franklin King, Madagascar
Assisted by Monika Friedrich, Madagascar

ORTHOCLASE - $KAlSi_3O_8$ - *Vitreous colourless, white, pale yellow, pink, grey, etc., Hardness 6-6.5.; Locality: Arizona, USA*

Photography by Jim Hughes, Assisted by ♪ Melody ♫
Collection of ♪ Melody ♫, Applewood, CO, USA

OSMIRIDIUM - *(Tiny Crystal)*
Iridium and Osmium in varying
proportions; Metallic luster;
Locality; California, USA

Photography by Jim Hughes, Assisted by ♪ Melody ♫
Collection of Colorado School Of Mines, CO, USA

OSUMILITE - *Composition not*
available; Locality: Oregon, USA

Photography by Jim Hughes, Assisted by ♪ Melody ♫
Collection of Colorado School Of Mines, CO, USA

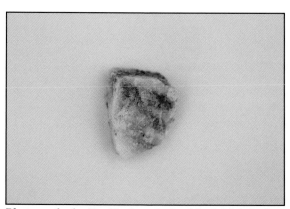

OWYHEEITE - *(With Friebergite)*
- $Ag_{3+x}Pb_{10-2x}Sb_{11+x}S_{28}$ - Metallic
grey, yellowish-white/grey;
Hardness 2.5; Locality: Nevada,
USA

Photography by Jim Hughes, Assisted by ♪ Melody ♫
Collection of ♪ Melody ♫, Applewood, CO, USA

Share your wisdom ♪

PACHNOLITE *(In Cryolite)* -
$NaCaAlF_6 \cdot H_2O$ - *Vitreous
colourless, white; Hardness 3;
Locality: Ivigtut, Greenland*

*Photography by Jim Hughes, Assisted by ♪ Melody ♫
Collection of ♪ Melody ♫, Applewood, CO, USA*

PALERMOITE -
$(Li,Na)_2(Sr,Ca)Al_4(PO_4)_4(OH)_4$ -
*Vitreous/sub-adamantine
colourless, white to yellow;
Hardness 5.5; Locality: New
Hampshire, USA*

*Photography by Jim Hughes, Assisted by ♪ Melody ♫
Collection of ♪ Melody ♫, Applewood, CO, USA*

PALERMOITE -
$(Li,Na)_2(Sr,Ca)Al_4(PO_4)_4(OH)_4$ -
*Vitreous/sub-adamantine
colourless, white to yellow;
Hardness 5.5; Locality: New
Hampshire, USA*

*Photography by Jim Hughes, Assisted by ♪ Melody ♫
Collection of ♪ Melody ♫, Applewood, CO, USA*

PAPAGOITE (In Quartz) - CuAlSi$_2$O$_6$(OH)$_6$ - Translucent blue;
Hardness 5-5.5; Locality: Messina, RSA, Africa
Photography by Jim Hughes, Assisted by ♪ Melody ♫
Collection of ♪ Melody ♫, Applewood, CO, USA

PAPAGOITE - *(In Quartz With Copper) - CaCuAlSi$_2$O$_6$(OH)$_6$ - Translucent blue; Hardness from 5-5.5; Locality: Messina, Republic Of South Africa, Africa*

Photography by Jim Hughes, Assisted by ♪ Melody ♫
Collection of ♪ Melody ♫, Applewood, CO, USA

PARISITE - *A fluocarbonate of cerium, [(Ce,La,Di)F]$_2$Ca (CO$_3$)$_3$ -Brown to yellow colour; Hardness 4.5; Locality: Montana, USA*

Photography by Jim Hughes, Assisted by ♪ Melody ♫
Collection of ♪ Melody ♫, Applewood, CO, USA

PASCOITE - *Ca$_3$V$_{10}$O$_{28}$♥17H$_2$O - Vitreous/sub-adamantine red/yellow/orange; Hardness 2.5; Locality: Colorado*

Photography by Jim Hughes, Assisted by ♪ Melody ♫
Collection of ♪ Melody ♫, Applewood, CO, USA

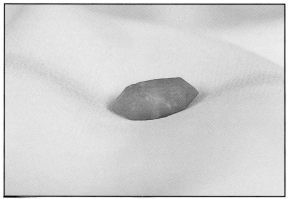

PECOS DIAMOND - SiO_2 with impurities - Small terminated crystal with lustrous light to deep peach colour; Hardness 7; Locality: Texas, USA

PECTOLITE - $NaCa_2Si_3O_8(OH)$ - Silky/sub-vitreous whitish, greyish; Hardness 5; Locality: Quebec, Canada

PENNINITE - $H_8(Mg,Fe)_5Al_2Si_3O_{18}$ - Vitreous to pearly/Light green, pink, red, yellow, violet, white; Hardness from 2-2.5; Locality: Vermont, USA

*PERICLASE - (With Chlorite) -
MgO - Vitreous colourless,
greyish-white, yellow, green,
black; Hardness 5.5; Locality:
Colorado, USA*

*Photography by Jim Hughes, Assisted by ♪ Melody ♫
Collection of ♪ Melody ♫, Applewood, CO, USA*

*PERIDOT - (Mg,Fe)₂SiO₄ -
Transparent vitreous pale green
to yellowish-green; Hardness
6.5-7; Locality: Arizona, USA*

*Photography by Jim Hughes, Assisted by ♪ Melody ♫
Collection of Julianne Guilbault, Lakewood, CO, USA*

*PEROVSKITE - CaTiO₃ -
Adamantine/metallic black,
brown, yellow; Hardness 5.5;
Locality: California, USA*

*Photography by Jim Hughes, Assisted by ♪ Melody ♫
Collection of ♪ Melody ♫, Applewood, CO, USA*

PETOSKEY STONE - *Fossilized hexagonaria coral from Michigan.*

Photography by Jim Hughes, Assisted by ♪ Melody ♫
Collection of ♪ Melody ♫, Applewood, CO, USA

PETRIFIED WOOD - *Ancient wood replaced and hardened by another mineral; Hardness variable; Locality: Wyoming, USA*

Photography by Jim Hughes, Assisted by ♪ Melody ♫
Collection of ♪ Melody ♫, Applewood, CO, USA

PETRIFIED BOG - *Zimbabwe*

Photography by Jim Hughes, Assisted by ♪ Melody ♫
Collection of ♪ Melody ♫, Applewood, CO, USA

Petrified Wood - Arizona, USA
Photography by Jim Hughes, Assisted by ♪ Melody ♫
Collection of ♪ Melody ♫, Applewood, CO, USA

♥ *LOVE IS IN THE EARTH* ♥

PETZITE - *(Silvery with Yellow Altaite)- Ag_3AuTe_2 - Metallic grey; Hardness 2.5-3; Locality: Colorado, USA*

Photography by Jim Hughes, Assisted by ♪ Melody
Collection of ♪ Melody ♫, Applewood, CO, USA

PHENACITE - Be₂SiO₄ - Vitreous colourless, yellow, red, brown; Hardness 7.5-8; Locality: Specimen on left, from Zimbabwe, Africa; Specimen on right from Minas, Gerais, Brasil

Photography by Jim Hughes, Assisted by ♪ Melody ♫
Collection of ♪ Melody ♫, Applewood, CO, USA

PHENACITE - Be₂SiO₄ - Vitreous colourless, yellow, red, brown; shown in matrix; Hardness 7.5-8; Locality: Russia

Photography by Jim Hughes, Assisted by ♪ Melody ♫
Collection of ♪ Melody ♫, Applewood, CO, USA
Gift of Ed Maslovicz, Sanctuary Crystals, Alsip, IL, USA

PHENACITE - Be₂SiO₄ - Vitreous colourless, yellow, red, brown; shown in matrix; Hardness 7.5-8; Locality: Russia

Photography by Jim Hughes, Assisted by ♪ Melody ♫
Collection of ♪ Melody ♫, Applewood, CO, USA

PHENACITE - Be_2SiO_4 -
*Vitreous <u>colourless</u>, yellow, red,
brown; Hardness 7.5-8;
Locality: Madagascar*

Photography by Jim Hughes, Assisted by ♪ Melody ♫
Collection of ♪ Melody ♫, Applewood, CO, USA

PHILLIPSITE -
$K(Ca_{0.5},Na)_2(Si_5Al_3)O_{16}$♥$6H_2O$ -
*Vitreous white, <u>reddish</u>,
yellowish; Hardness 4-4.5;
Locality: Arizona, USA*

Photography by Jim Hughes, Assisted by ♪ Melody ♫
Collection of ♪ Melody ♫, Applewood, CO, USA

PHLOGOPITE -
$H_2KMg_3Al(SiO_4)_3$ - *Yellow-
brown to brown-red, green,
white, colourless, sometimes
with copper-like reflection and
asterism; Pleochroistic;
Hardness 2.5-3; Locality: Kola,
Kovdor, Russia*

Photography by Jim Hughes, Assisted by ♪ Melody ♫
Collection of ♪ Melody ♫, Applewood, CO, USA

PHOSPHORITE - Composition similar to Apatite, but also contains fibrous concretionary and may be partially scaly; Hardness: 4.5-5; Locality: Germany

Photography by Jim Hughes, Assisted by ♪ Melody ♫
Collection of ♪ Melody ♫, Applewood, CO, USA
Gift from Ms Sigari, Germany

PICASSO STONE - Composition is a type of marble with mineral inclusions; Locality unknown to author.

Photography by Jim Hughes, Assisted by ♪ Melody ♫
Collection of ♪ Melody ♫, Applewood, CO, USA

PICASSO STONE - Composition is a type of marble with mineral inclusions; Locality unknown to author.

Photography by Jim Hughes, Assisted by ♪ Melody ♫
Collection of ♪ Melody ♫, Applewood, CO, USA

PICROLITE - $H_4Mg_3Si_2O_9$ - A variance of Serpentine, fibrous and columnar, with colour deep green, grey, brown; Hardness: 2.5-4; Locality: Ontario, Canada

*Photography by Jim Hughes, Assisted by ♪ Melody ♫
Collection of ♪ Melody ♫, Applewood, CO, USA*

PIETERSITE - SiO_2 with inclusions of pseudomorphs after asbestos; Colour blue/black with chatoyancy; Hardness 7; Namibia, Africa; Discovered by Sid Pieters, Windhoek, Namibia, Africa.

*Photography by Jim Hughes, Assisted by ♪ Melody ♫
Collection of Bob Jackson, Applewood, CO, USA*

PIETERSITE - SiO_2 with inclusions of pseudomorphs after asbestos; Colour blue/black with chatoyancy; Hardness 7; Namibia, Africa; Discovered by Sid Pieters, Windhoek, Namibia, Africa.

*Photography by Jim Hughes, Assisted by ♪ Melody ♫
Collection of Bob Jackson, Applewood, CO, USA*

PINAKIOLITE - $Mg_2MnO_2(BO_3)$ -
Metallic black; Hardness 6;
Varmland, Sweden

Photography by Jim Hughes, Assisted by ♪ Melody ♫
Collection of Colorado School Of Mines, CO, USA

PIPESTONE - *Composition*
variable aluminum silicate with
iron impurities; Locality: Arizona,
USA

Photography by Jim Hughes, Assisted by ♪ Melody ♫
Collection of ♪ Melody ♫, Applewood, CO, USA

PITCHSTONE - *Volcanic glass*
which is dull and pitch-like in
luster; Hardness 5-5.5; Locality;
India

Photography by Jim Hughes, Assisted by ♪ Melody ♫
Collection of ♪ Melody ♫, Applewood, CO, USA

PLANCHEITE -
$Cu_8(Si_4O_{11})_2(OH)_4 \heartsuit H_2O$ -
Translucent dark blue in
Chrysocolla; Hardness 5.5;
Locality: Arizona, USA

Photography by Jim Hughes, Assisted by ♪ Melody ♫
Collection of ♪ Melody ♫, Applewood, CO, USA

PLATINUM (With Palladium) -
Pt with Pd - Metallic white/grey;
Hardness 4.5-5; Locality:
Alaska, USA

Photography by Jim Hughes, Assisted by ♪ Melody ♫
Collection of ♪ Melody ♫, Applewood, CO, USA

PLATINUM - (With Gold in
Quartz) - Locality: Alaska, USA

Photography by Jim Hughes, Assisted by ♪ Melody ♫
Collection of Bob Jackson, Applewood, CO, USA

POWELLITE - $CaMoO_4$ - Sub-adamantine/lustrous <u>yellow</u>, brown, grey, blue, etc.; Hardness 3.5-4; Locality: India

Photography by Jim Hughes, Assisted by ♪ Melody ♫
Collection of ♪ Melody ♫, Applewood, CO, USA

PREHNITE - $Ca_2Al(Si,Al)_4O_{10}(OH)_2$ - Vitreous/pearly light <u>green</u>, white, grey; Hardness 6-6.5; Locality: Cape of Good Hope, Republic Of South Africa, Africa

Photography by Jim Hughes, Assisted by ♪ Melody ♫
Collection of ♪ Melody ♫, Applewood, CO, USA

PROUSTITE - Ag_3AsS_3 - (Known Also As Ruby Silver) Adamantine translucent red when lit, grey metallic in normal lighting; Hardness 2-2.5; Locality: Mexico

Photography by Jim Hughes, Assisted by ♪ Melody ♫
Collection of ♪ Melody ♫, Applewood, CO, USA
Gift from Angel Torrecillas, Mexico

PROUSTITE - Ag₃AsS₃ - (Known Also As Ruby Silver) Adamantine translucent red when lit, grey metallic in normal lighting; Hardness 2-2.5; Locality: Mexico; Photograph with back-light.

Photography by Jim Hughes, Assisted by ♪ Melody ♫
Collection of ♪ Melody ♫, Applewood, CO, USA
Gift of Angel Torrecillas, Mexico

PSILOMELANE - MnO₂ Colloidal with impurities - Opaque sub-metallic to dull iron-black to steel-grey; Hardness 5-7; Locality: Mexico

Photography by Jim Hughes, Assisted by ♪ Melody ♫
Collection of Bob Jackson, Applewood, CO, USA

PSILOMELANE - MnO₂ - (Drusy); Colloidal with impurities - Opaque sub-metallic to dull iron-black to steel-grey; Hardness 5-7; Locality: Mexico

Photography by Jim Hughes, Assisted by ♪ Melody ♫
Collection of Bob Jackson, Applewood, CO, USA

PUMICE - *Fine-grained volcanic rock with well-defined cellular structure; Hardness variable; Locality: Hawaii, USA*

Photography by Jim Hughes, Assisted by ♪ Melody ♫
Collection of ♪ Melody ♫, Applewood, CO, USA

PUMPELLYITE (Green) - (With Dickite Crystals); $6CaO \heartsuit 3Al_2O_3 \heartsuit 7SiO_2 \heartsuit 4H_{20}$ - *Fibrous or plates of blue-green; Hardness: 5.5; Locality: Czechoslovakia*

Photography by Jim Hughes, Assisted by ♪ Melody ♫
Collection of ♪ Melody ♫, Applewood, CO, USA

PURPURITE - $(Mn,Fe)PO_4$ - *Satiny deep rose, reddish-purple, dark brown; Hardness: 4-4.5; Locality: Namibia, Africa*

Photography by Jim Hughes, Assisted by ♪ Melody ♫
Collection of Bob Jackson, Applewood, CO, USA

PYRITE (Over Quartz) - FeS₂ -
Metallic yellow; Hardness 6-6.5;
Locality: Mexico

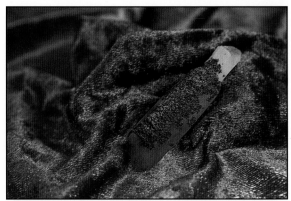

Photography by Jim Hughes, Assisted by ♪ Melody ♫
Collection of ♪ Melody ♫, Applewood, CO, USA

PYRITE (Cast) - FeS₂ - Metallic
yellow; Hardness 6-6.5;
Locality: Missouri, USA

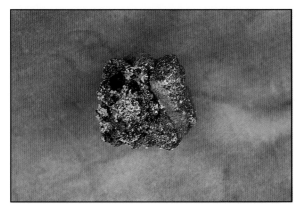

Photography by Jim Hughes, Assisted by ♪ Melody ♫
Collection of ♪ Melody ♫, Applewood, CO, USA

PYRITE (With Marcasite) -
FeS₂ with Marcasite - Metallic
yellow; Hardness 6-6.5;
Locality: Mexico

Photography by Jim Hughes, Assisted by ♪ Melody ♫
Collection of ♪ Melody ♫, Applewood, CO, USA

PYRITE (With Quartz) - FeS₂ -
Metallic yellow; Hardness 6-6.5;
Locality: Washington, USA

Photography by Jim Hughes, Assisted by ♪ Melody ♫
Collection of ♪ Melody ♫, Applewood, CO, USA

PYRITE (Arseno) - FeAsS-
Metallic yellow; Hardness 6-6.5;
Locality: New Jersey, USA

Photography by Jim Hughes, Assisted by ♪ Melody ♫
Collection of Bob Jackson, Applewood, CO, USA

PYRITE (In Quartz) - FeS₂ -
Metallic yellow; Hardness 6-6.5;
Locality: Switzerland

Photography by Jim Hughes, Assisted by ♪ Melody ♫
Collection of ♪ Melody ♫, Applewood, CO, USA

PYROLUSITE - Commonly *pseudomorphous after Manganite; Iron-black to steel-grey, sometimes bluish; submetallic and opaque; Hardness 2-2.5; Locality: Michigan*

Photography by Jim Hughes, Assisted by ♪ Melody ♫
Collection of Adria Parker, Kennewick WA, USA

PYROMORPHITE -
$Pb_5(PO_4)_3Cl$ - *Resinous sub-adamantine* <u>green</u>, *yellow, brown, etc.; Hardness 3.5-4; Locality: Queensland, Australia*

Photography by Jim Hughes, Assisted by ♪ Melody ♫
Collection of ♪ Melody ♫, Applewood, CO, USA

PYROPE - $Mg_3Al_2(SiO_4)_3$ - *Vitreous/resinous deep red; Hardness 6.5-7.5; Locality: Arizona, USA*

Photography by Jim Hughes, Assisted by ♪ Melody ♫
Collection of ♪ Melody ♫, Applewood, CO, USA

QUARTZ - alpha-SiO$_2$ - Vitreous; Hardness 7; Configuration: ACTIVATION (Left) QUARTZ CRYSTAL; Location: Minas Gerais, Brasil

*Photography by Jim Hughes, Assisted by ♪ Melody ♫
Collection of ♪ Melody ♫, Applewood, CO, USA*

QUARTZ - alpha-SiO$_2$ - Vitreous; Hardness 7; Configuration: ACTIVATION (Right) QUARTZ CRYSTAL; Location: Minas Gerais, Brasil

*Photography by Jim Hughes, Assisted by ♪ Melody ♫
Collection of ♪ Melody ♫, Applewood, CO, USA*

QUARTZ - alpha-SiO$_2$ - Vitreous; Hardness 7; Configuration: APERTURE QUARTZ CRYSTAL; Location: Minas Gerais, Brasil

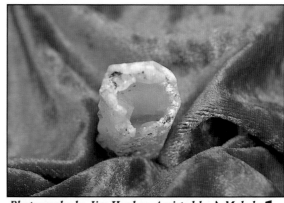

*Photography by Jim Hughes, Assisted by ♪ Melody ♫
Collection of ♪ Melody ♫, Applewood, CO, USA*

QUARTZ - alpha-SiO$_2$ - Vitreous; Hardness 7; Configuration: BARNACLE (Laser Wand) QUARTZ CRYSTAL; Location: Minas Gerais, Brasil

Photography by Jim Hughes, Assisted by ♪ Melody ♫
Collection of ♪ Melody ♫, Applewood, CO, USA

QUARTZ - alpha-SiO$_2$ - Vitreous; Hardness 7; Configuration: BRIDGE QUARTZ CRYSTAL; Location: Arkansas, USA

Photography by Jim Hughes, Assisted by ♪ Melody ♫
Collection of ♪ Melody ♫, Applewood, CO, USA

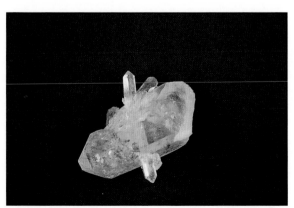

QUARTZ - alpha-SiO$_2$ - Vitreous; Hardness 7; Configuration: BRIDGE QUARTZ CRYSTAL; Location: Arkansas, USA

Photography by Jim Hughes, Assisted by ♪ Melody ♫
Collection of ♪ Melody ♫, Applewood, CO, USA

QUARTZ - alpha-SiO$_2$ - Vitreous; Hardness 7; Configuration: BRIDGE QUARTZ CRYSTAL (Many Bridges); Location: Arkansas, USA

Photography by Jim Hughes, Assisted by ♪ Melody ♫
Collection of ♪ Melody ♫, Applewood, CO, USA

QUARTZ - alpha-SiO$_2$ - Vitreous; Hardness 7; Configuration: CARVED (Natural) QUARTZ CRYSTAL; Location: Arkansas, USA

Photography by Jim Hughes, Assisted by ♪ Melody ♫
Collection of ♪ Melody ♫, Applewood, CO, USA
Gift of Betty & Wayne Green, Sedona Crystal Mine, AZ

QUARTZ - alpha-SiO$_2$ - Vitreous; Hardness 7; Configuration: CHANNELER/ RECORD KEEPER QUARTZ CRYSTAL; Location: Minas Gerais, Brasil

Photography by Jim Hughes, Assisted by ♪ Melody ♫
Collection of ♪ Melody ♫, Applewood, CO, USA

QUARTZ - alpha-SiO$_2$ - Vitreous; Hardness 7; Configuration: CLUSTER QUARTZ CRYSTALS; Location: Peru

Photography by Jim Hughes, Assisted by ♪ Melody ♫
Collection of ♪ Melody ♫, Applewood, CO, USA

QUARTZ - alpha-SiO$_2$ - Vitreous; Hardness 7; Configuration: CLUSTER QUARTZ CRYSTALS; Location: Arkansas, USA

Photography by Jim Hughes, Assisted by ♪ Melody ♫
Collection of ♪ Melody ♫, Applewood, CO, USA

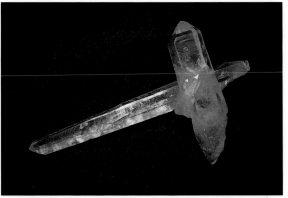

QUARTZ - alpha-SiO$_2$ - Vitreous; Hardness 7; Configuration: CROSS QUARTZ CRYSTAL; Location: Minas Gerais, Brasil

Photography by Jim Hughes, Assisted by ♪ Melody ♫
Collection of ♪ Melody ♫, Applewood, CO, USA

QUARTZ - alpha-SiO₂ - Vitreous; Hardness 7; Configuration: CURVED QUARTZ CRYSTAL; Location: Minas Gerais, Brasil

Photography by Jim Hughes, Assisted by ♪ Melody ♫
Collection of ♪ Melody ♫, Applewood, CO, USA

QUARTZ - alpha-SiO₂ - Vitreous; Hardness 7; Configuration: CURVED QUARTZ CRYSTAL; Location: Minas Gerais, Brasil

Photography by Jim Hughes, Assisted by ♪ Melody ♫
Collection of ♪ Melody ♫, Applewood, CO, USA

QUARTZ - alpha-SiO₂ - Vitreous; Hardness 7; Configuration: DOUBLE TERMINATED QUARTZ CRYSTAL; Location: Minas Gerais, Brasil

Photography by Jim Hughes, Assisted by ♪ Melody ♫
Collection of ♪ Melody ♫, Applewood, CO, USA

QUARTZ - alpha-SiO$_2$ - Vitreous; Hardness 7; Configuration: DOUBLE TERMINATED QUARTZ CRYSTAL; Location: Arkansas, USA

Photography by Jim Hughes, Assisted by ♪ Melody ♫
Collection of ♪ Melody ♫, Applewood, CO, USA

QUARTZ - alpha-SiO$_2$ - Vitreous; Hardness 7; Configuration: ELESTIAL (Smokey Quartz [organic impurities or natural exposure to radioactivity causing the colour]) CRYSTAL; Location: Minas Gerais, Brasil

Photography by Franklin King, Madagascar
Assisted by Monika Friedrich, Madagascar

QUARTZ - alpha-SiO$_2$ - Vitreous; Hardness 7; Configuration: ELESTIAL (Smokey Quartz) CRYSTAL; Location: Bahia, Brasil

Photography by Jim Hughes, Assisted by ♪ Melody ♫
Collection of ♪ Melody ♫, Applewood, CO, USA

QUARTZ - alpha-SiO$_2$ - Vitreous; Hardness 7; Configuration: ETCHED QUARTZ CRYSTAL; Location: Arkansas, USA

Photography by Jim Hughes, Assisted by ♪ Melody ♫
Collection of ♪ Melody ♫, Applewood, CO, USA

QUARTZ - alpha-SiO$_2$ - Vitreous; Hardness 7; Configuration: FADEN QUARTZ CRYSTAL; Location: Minas Gerais, Brasil

Photography by Jim Hughes, Assisted by ♪ Melody ♫
Collection of ♪ Melody ♫, Applewood, CO, USA
Gift of Jose Orizon de Almeida, Belo Horizonte,
Minas Gerais, Brasil

QUARTZ - alpha-SiO$_2$ - Vitreous; Hardness 7; Configuration: FADEN QUARTZ CRYSTAL; Location: Arkansas, USA

Photography by Jim Hughes, Assisted by ♪ Melody ♫
Collection of ♪ Melody ♫, Applewood, CO, USA

QUARTZ - alpha-SiO$_2$ - Vitreous; Hardness 7; Configuration: GENERATOR (Chevron Amethyst Polished) CRYSTAL; Location: India

Photography by Jim Hughes, Assisted by ♪ Melody ♫
Collection of ♪ Melody ♫, Applewood, CO, USA

QUARTZ - alpha-SiO$_2$ - Vitreous; Hardness 7; Configuration: HORN OF PLENTY QUARTZ CRYSTAL; Location: Minas Gerais, Brasil

Photography by Jim Hughes, Assisted by ♪ Melody ♫
Collection of ♪ Melody ♫, Applewood, CO, USA

QUARTZ - alpha-SiO$_2$ - Vitreous; Hardness 7; Configuration: JAPAN LAW TWIN QUARTZ CRYSTAL; Location: California, USA

Photography by Jim Hughes, Assisted by ♪ Melody ♫
Collection of ♪ Melody ♫, Applewood, CO, USA

QUARTZ - alpha-SiO$_2$ - Vitreous; Hardness 7; Configuration:
JACARE QUARTZ CRYSTAL CLUSTER
(With Inclusions); Location: Minas Gerais, Brasil
Photography by Jim Hughes, Assisted by ♪ Melody ♫
Collection of Jackson & ♪ Melody ♫, Applewood, CO, USA

QUARTZ - alpha-SiO$_2$ - Vitreous; Hardness 7; Configuration: KEY (ELESTIAL SMOKEY QUARTZ) CRYSTAL; Location: Bahia, Brasil

Photography by Jim Hughes, Assisted by ♪ Melody ♫
Collection of Jackson & Melody, Applewood, CO, USA

QUARTZ - alpha-SiO$_2$ - Vitreous; Hardness 7; Configuration: KEY (Double Terminated) QUARTZ CRYSTAL; Location: Minas Gerais, Brasil

Photography by Jim Hughes, Assisted by ♪ Melody ♫
Collection of ♪ Melody ♫, Applewood, CO, USA

QUARTZ - alpha-SiO$_2$ - Vitreous; Hardness 7; Configuration: LASER WAND (Chimney) QUARTZ CRYSTAL; Location: Minas Gerais, Brasil

Photography by Jim Hughes, Assisted by ♪ Melody ♫
Collection of ♪ Melody ♫, Applewood, CO, USA

QUARTZ - alpha-SiO$_2$ - Vitreous; Hardness 7; Configuration: LASER WAND QUARTZ CRYSTAL CLUSTER; Location: Minas Gerais, Brasil

Photography by Jim Hughes, Assisted by ♪ Melody ♫
Collection of Julianne Guilbault, Lakewood, CO, USA

QUARTZ - alpha-SiO$_2$ - Vitreous; Hardness 7; Configuration: LASER WAND QUARTZ CRYSTAL (Triplet); Location: Minas Gerais, Brasil

Photography by Jim Hughes, Assisted by ♪ Melody ♫
Collection of ♪ Melody ♫, Applewood, CO, USA

QUARTZ - alpha-SiO$_2$ - Vitreous; Hardness 7; Configuration: MANIFESTATION QUARTZ CRYSTAL; Location: Arkansas, USA

Photography by Jim Hughes, Assisted by ♪ Melody ♫
Collection of ♪ Melody ♫, Applewood, CO, USA

QUARTZ - alpha-SiO₂ - Vitreous; Hardness 7; Configuration:
LASER WAND MANIFESTATION QUARTZ CRYSTAL
Location: Minas Gerais, Brasil
Photography by Jim Hughes, Assisted by ♪ Melody ♫
Collection of Bob Jackson & Melody, Applewood, CO, USA

QUARTZ - alpha-SiO₂ - Vitreous; Hardness 7; Configuration: NODULE QUARTZ CRYSTAL; Location: Minas Gerais, Brasil

Photography by Jim Hughes, Assisted by ♪ Melody ♫
Collection of ♪ Melody ♫, Applewood, CO, USA

QUARTZ - alpha-SiO₂ - Vitreous; Hardness 7; Configuration: OBELISK (Tourmalinated - Polished) QUARTZ; Location: Minas Gerais, Brasil

Photography by Jim Hughes, Assisted by ♪ Melody ♫
Collection of ♪ Melody ♫, Applewood, CO, USA

QUARTZ - alpha-SiO₂ - Vitreous; Hardness 7; Configuration: PINNACLE QUARTZ CRYSTAL; Location: USA

Photography by Jim Hughes, Assisted by ♪ Melody ♫
Collection of ♪ Melody ♫, Applewood, CO, USA
Gift of Amy Kiesbuy, Idaho, USA

QUARTZ - alpha-SiO$_2$ - Vitreous;
Hardness 7; Configuration:
POCKET QUARTZ CRYSTAL;
Location: Minas Gerais, Brasil

Photography by Jim Hughes, Assisted by ♪ Melody ♫
Collection of ♪ Melody ♫, Applewood, CO, USA
Gift from Elenita Oliveira, Belo Horizonte,
Minas Gerais, Brasil

QUARTZ - alpha-SiO$_2$ - Vitreous;
Hardness 7; Configuration:
RAINBOW QUARTZ (Smokey)
CRYSTAL; Location: Montana,
USA

Photography by Jim Hughes, Assisted by ♪ Melody ♫
Collection of Bob Jackson, Applewood, CO, USA

QUARTZ - alpha-SiO$_2$ - Vitreous;
Hardness 7; Configuration:
RECORD KEEPER QUARTZ
CRYSTAL; Location: Minas
Gerais, Brasil

Photography by Jim Hughes, Assisted by ♪ Melody ♫
Collection of ♪ Melody ♫, Applewood, CO, USA

QUARTZ - alpha-SiO$_2$ - Vitreous; Hardness 7; Configuration: SCEPTOR (Smokey) QUARTZ CRYSTAL; Location: Montana, USA

Photography by Jim Hughes, Assisted by ♪ Melody ♫
Collection of ♪ Melody ♫, Applewood, CO, USA

QUARTZ - alpha-SiO$_2$ - Vitreous; Hardness 7; Configuration: SCEPTOR QUARTZ CRYSTAL; Location: Arkansas, USA

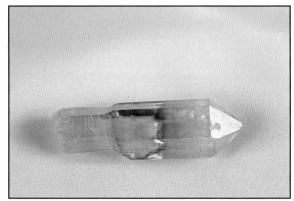

Photography by Jim Hughes, Assisted by ♪ Melody ♫
Collection of ♪ Melody ♫, Applewood, CO, USA

QUARTZ - alpha-SiO$_2$ - Vitreous; Hardness 7; Configuration: SEER STONE QUARTZ; Location: Minas Gerais, Brasil

Photography by Jim Hughes, Assisted by ♪ Melody ♫
Collection of ♪ Melody ♫, Applewood, CO, USA

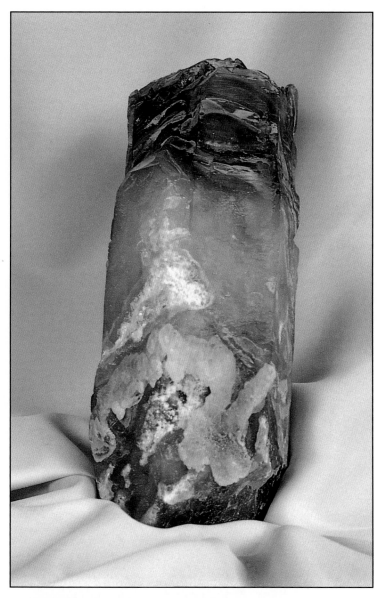

QUARTZ - alpha-SiO$_2$ - Vitreous; Hardness 7; Configuration:
SELF-HEALED ROSE, AND SMOKEY ELESTIATED SCEPTOR QUARTZ CRYSTAL
Location: Bahia, Brasil
Photography by Jim Hughes, Assisted by ♪ Melody ♫
Collection of Jackson & Melody, Applewood, CO, USA

QUARTZ - alpha-SiO$_2$ - Vitreous; Hardness 7; Configuration: SELF HEALED QUARTZ CRYSTAL (Elestiated Base); Location: Messina, Republic of South Africa, Africa

Photography by Jim Hughes, Assisted by ♪ Melody ♫
Collection of ♪ Melody ♫, Applewood, CO, USA

QUARTZ - alpha-SiO$_2$ - Vitreous; Hardness 7; Configuration: SHEET QUARTZ; Location: Minas Gerais, Brasil

Photography by Jim Hughes, Assisted by ♪ Melody ♫
Collection of ♪ Melody ♫, Applewood, CO, USA

QUARTZ - alpha-SiO$_2$ - Vitreous; Hardness 7; Configuration: SHOVEL QUARTZ; Location: Minas Gerais, Brasil

Photography by Jim Hughes, Assisted by ♪ Melody ♫
Collection of ♪ Melody ♫, Applewood, CO, USA

QUARTZ - alpha-SiO₂ - Vitreous;
Hardness 7; Configuration:
SHOVEL KEY QUARTZ
CRYSTAL; Location: Minas
Gerais, Brasil

Photography by Jim Hughes, Assisted by ♪ Melody ♫
Collection of ♪ Melody ♫, Applewood, CO, USA

QUARTZ - alpha-SiO₂ - Vitreous;
Hardness 7; Configuration:
SINGING (Double Terminated)
QUARTZ CRYSTAL; Location:
Minas Gerais, Brasil

Photography by Jim Hughes, Assisted by ♪ Melody ♫
Collection of ♪ Melody ♫, Applewood, CO, USA

QUARTZ - alpha-SiO₂ - Vitreous;
Hardness 7; Configuration:
SPADE QUARTZ; Location:
Minas Gerais, Brasil

Photography by Jim Hughes, Assisted by ♪ Melody ♫
Collection of ♪ Melody ♫, Applewood, CO, USA

QUARTZ - alpha-SiO$_2$ - Vitreous; Hardness 7; Configuration: RAINBOW and SPHERE (Polished Quartz); Location: Minas Gerais, Brasil

Photography by Jim Hughes, Assisted by ♪ Melody ♫
Collection of ♪ Melody ♫, Applewood, CO, USA

QUARTZ - alpha-SiO$_2$ - Vitreous; Hardness 7; Configuration: SPIRAL QUARTZ CRYSTAL; Location: Minas Gerais, Brasil

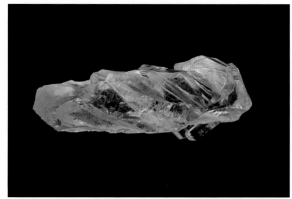

Photography by Jim Hughes, Assisted by ♪ Melody ♫
Collection of ♪ Melody ♫, Applewood, CO, USA

QUARTZ - alpha-SiO$_2$ - Vitreous; Hardness 7; Configuration: SPIRAL SINGING QUARTZ CRYSTALS; Location: Minas Gerais, Brasil

Photography by Jim Hughes, Assisted by ♪ Melody ♫
Collection of ♪ Melody ♫, Applewood, CO, USA

QUARTZ - alpha-SiO$_2$ - Vitreous; Hardness 7; Configuration: TABBY QUARTZ CRYSTAL; Location: Minas Gerais, Brasil

Photography by Jim Hughes, Assisted by ♪ Melody ♫
Collection of ♪ Melody ♫, Applewood, CO, USA

QUARTZ - alpha-SiO$_2$ - Vitreous; Hardness 7; Configuration: TIBETAN QUARTZ CLUSTER; Location: Tibet

Photography by Jim Hughes, Assisted by ♪ Melody ♫
Collection of ♪ Melody ♫, Applewood, CO, USA

QUARTZ - alpha-SiO$_2$ - Vitreous; Hardness 7; Configuration: TWIN QUARTZ CRYSTALS; Location: Arkansas, USA

Photography by Jim Hughes, Assisted by ♪ Melody ♫
Collection of ♪ Melody ♫, Applewood, CO, USA

QUARTZ - (Specialty) AQUA AURA QUARTZ (Faceted) - SiO₂ with atomic nuclei of gold;- Vitreous; Hardness 7; Location: Minas Gerais, Brasil

Photography by Jim Hughes, Assisted by ♪ Melody ♫
Collection of ♪ Melody ♫, Applewood, CO, USA

QUARTZ - (Specialty) AQUA AURA QUARTZ (Crystal) - SiO₂ with atomic nuclei of gold;- Vitreous; Hardness 7; Location: Minas Gerais, Brasil

Photography by Jim Hughes, Assisted by ♪ Melody ♫
Collection of ♪ Melody ♫, Applewood, CO, USA

QUARTZ - (Specialty) ASBESTOS IN QUARTZ - SiO₂ - Vitreous; Hardness 7; Location: Minas Gerais, Brasil

Photography by Jim Hughes, Assisted by ♪ Melody ♫
Collection of ♪ Melody ♫, Applewood, CO, USA

QUARTZ - (Specialty)
ASBESTOS IN QUARTZ - SiO$_2$ -
Vitreous; Hardness 7; Location:
Minas Gerais, Brasil

Photography by Jim Hughes, Assisted by ♪ Melody ♫
Collection of ♪ Melody ♫, Applewood, CO, USA

QUARTZ - (Specialty)
ASTERATED QUARTZ - SiO$_2$ -
Vitreous; Hardness 7; Location:
Minas Gerais, Brasil

Photography by Jim Hughes, Assisted by ♪ Melody ♫
Collection of ♪ Melody ♫, Applewood, CO, USA
Gift of Rackel and Natalino Eugenio Oliveira

QUARTZ - (Specialty) CITRINE
AND QUARTZ - SiO$_2$ - Vitreous;
Hardness 7; Location: Minas
Gerais, Brasil

Photography by Jim Hughes, Assisted by ♪ Melody ♫
Collection of ♪ Melody ♫, Applewood, CO, USA

QUARTZ - (Specialty) COLOURED (BLUE) QUARTZ - SiO$_2$ with tiny rutile, tourmaline or zoisite inclusions- Vitreous; Hardness 7; Location: Minas Gerais, Brasil

Photography by Jim Hughes, Assisted by ♪ Melody ♫
Collection of ♪ Melody ♫, Applewood, CO, USA

QUARTZ - (Specialty) COLOURED (GREEN) QUARTZ - SiO$_2$ with inclusions (Composition currently undefined)- Vitreous; Hardness of 7; Location: Madagascar

Photography by Jim Hughes, Assisted by ♪ Melody ♫
Collection of ♪ Melody ♫, Applewood, CO, USA
Gift of Ken Harsh, Ohio, USA

QUARTZ - (Specialty) COLOURED (LAVENDER) QUARTZ - SiO$_2$ with Manganese/Titanium - Vitreous; Hardness 7; Location: Minas Gerais, Brasil

Photography by Jim Hughes, Assisted by ♪ Melody ♫
Collection of ♪ Melody ♫, Applewood, CO, USA

QUARTZ - (Specialty)
COLOURED (LAVENDER)
QUARTZ - SiO$_2$ with
Manganese/Titanium - Vitreous;
Hardness 7; Location: Minas
Gerais, Brasil

Photography by Jim Hughes, Assisted by ♪ Melody ♫
Collection of ♪ Melody ♫, Applewood, CO, USA

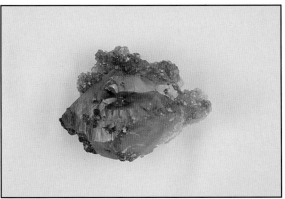

QUARTZ - (Specialty)
COLOURED (RED) QUARTZ -
SiO$_2$ (Composition currently
undefined - speculation of SiO$_2$
with Iron and/or Hematite -
Vitreous; Hardness 7; Location:
Canada

Photography by Jim Hughes, Assisted by ♪ Melody ♫
Collection of ♪ Melody ♫, Applewood, CO, USA

QUARTZ - (Specialty)
COLOURED (RED) QUARTZ -
SiO$_2$ (Composition currently
undefined - speculation of SiO$_2$
with Iron and/or Hematite -
Vitreous; Hardness 7; Location:
Minas Gerais, Brasil

Photography by Jim Hughes, Assisted by ♪ Melody ♫
Collection of ♪ Melody ♫, Applewood, CO, USA

QUARTZ - (Specialty) COLOURED (STRAWBERRY) QUARTZ - SiO₂ (Composition currently undefined - speculation of SiO₂ with Ferric Iron, Rutile, and/or Lepidochrosite- Vitreous; Hardness 7; Location: Mexico

Photography by Jim Hughes, Assisted by ♪ Melody ♫
Collection of ♪ Melody ♫, Applewood, CO, USA

QUARTZ - (Specialty) COLOURED (STRAWBERRY) QUARTZ - SiO₂ (Composition currently undefined - speculation of SiO₂ with Ferric Iron, Rutile, and/or Lepidochrosite- Vitreous; Hardness 7; Location: Mexico

Photography by Jim Hughes, Assisted by ♪ Melody ♫
Collection of ♪ Melody ♫, Applewood, CO, USA

QUARTZ - (Specialty) (WITH CUPRITE) - SiO₂ with Cuprite - Vitreous; Hardness 7; Location: Republic of South Africa, Africa

Photography by Jim Hughes, Assisted by ♪ Melody ♫
Collection of ♪ Melody ♫, Applewood, CO, USA

QUARTZ - (Specialty)
DENDRITIC QUARTZ - SiO$_2$
with Manganese Dioxide
inclusions - Vitreous; Hardness 7;
Location: Minas Gerais, Brasil

Photography by Jim Hughes, Assisted by ♪ Melody ♫
Collection of ♪ Melody ♫, Applewood, CO, USA

QUARTZ - (Specialty)
DENDRITIC QUARTZ - SiO$_2$
with Manganese Dioxide
inclusions - Vitreous; Hardness 7;
Location: Minas Gerais, Brasil

Photography by Jim Hughes, Assisted by ♪ Melody ♫
Collection of ♪ Melody ♫, Applewood, CO, USA

QUARTZ - (Specialty) GOLDEN
HEALER QUARTZ - SiO$_2$ with
impurities (possibly iron hydrates)
- Vitreous; Hardness 7; Location:
Unknown

Photography by Jim Hughes, Assisted by ♪ Melody ♫
Collection of ♪ Melody ♫, Applewood, CO, USA
Gift of "Crystal Dundee", USA

QUARTZ - (Specialty)
HARLEQUIN QUARTZ - SiO_2
with red dots (may be Hematite
or Lepidocrocite) - Vitreous;
Hardness 7; Location:
Madagascar

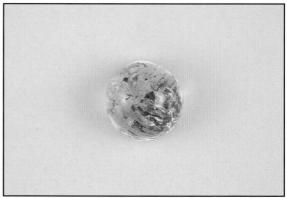

Photography by Jim Hughes, Assisted by ♪ Melody ♫
Collection of Bruce Camenetti, North Carolina, USA

QUARTZ - (Specialty)
HARLEQUIN QUARTZ - SiO_2
with red dots/strings (may be
Hematite or Lepidocrocite -
Vitreous; Hardness 7; Location:
Madagascar

Photography by Jim Hughes, Assisted by ♪ Melody ♫
Collection of ♪ Melody ♫, Applewood, CO, USA

QUARTZ - (Specialty)
HARLEQUIN QUARTZ - SiO_2
with red dots (may be Hematite
or Lepidocrocite - Vitreous;
Hardness 7; Location:
Madagascar

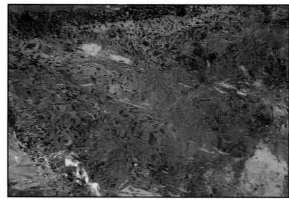

Photography by Jim Hughes, Assisted by ♪ Melody ♫
Collection of Bob Jackson, Applewood, CO, USA
(Gift of Monika Friedrich)

QUARTZ - (Included) [VOLCANO] - SiO$_2$ with undefined inclusions) - Vitreous; Hardness 7; Location: Minas Gerais, Brasil

Photography by Jim Hughes, Assisted by ♪ Melody ♫
Collection of ♪ Melody ♫, Applewood, CO, USA

QUARTZ - (Included) - SiO$_2$ with asbestos, etc.) - Vitreous; Hardness 7; Location: Minas Gerais, Brasil

Photography by Jim Hughes, Assisted by ♪ Melody ♫
Collection of ♪ Melody ♫, Applewood, CO, USA

QUARTZ - (Included) - SiO$_2$ with undefined inclusions) - Vitreous; Hardness 7; Location: Minas Gerais, Brasil

Photography by Jim Hughes, Assisted by ♪ Melody ♫
Collection of ♪ Melody ♫, Applewood, CO, USA

QUARTZ - (Included) - SiO₂ with undefined inclusions) - Vitreous; Hardness 7; Location: Zaire, Africa

$QUARTZ$ - (Included) - SiO_2 with undefined inclusions) - Vitreous; Hardness 7; Location: Zaire, Africa

Photography by Jim Hughes, Assisted by ♪ Melody ♫
Collection of Bob Jackson, Applewood, CO, USA

$QUARTZ$ - (Included) [Phantom] - SiO_2 with undefined inclusions) - Vitreous; Hardness 7; Location: Minas Gerais, Brasil

Photography by Jim Hughes, Assisted by ♪ Melody ♫
Collection of ♪ Melody ♫, Applewood, CO, USA

$QUARTZ$ - (Included Laser Wand) - SiO_2 with Limonite inclusions) - Vitreous; Hardness 7; Location: Minas Gerais, Brasil

Photography by Jim Hughes, Assisted by ♪ Melody ♫
Collection of ♪ Melody ♫, Applewood, CO, USA

QUARTZ - *(Included)* *[OPAL AURA QUARTZ] - SiO$_2$ with atomic nuclei of Platinum) - Vitreous; Hardness 7; Location: Minas Gerais, Brasil*

Photography by Jim Hughes, Assisted by ♪ Melody ♫
Collection of ♪ Melody ♫, Applewood, CO, USA
Gift of Richard Wegner, Wegner Quartz Crystal Mines, Mt. Ida, Arkansas, USA

QUARTZ - *(Included) - [BLACK PHANTOM] - SiO$_2$ with undefined black inclusions) - Vitreous; Hardness 7; Location: Minas Gerais, Brasil*

Photography by Jim Hughes, Assisted by ♪ Melody ♫
Collection of ♪ Melody ♫, Applewood, CO, USA

QUARTZ - *(Included) - [BLACK PHANTOM] - SiO$_2$ with undefined black inclusions) - Vitreous; Hardness 7; Location: Minas Gerais, Brasil*

Photography by Jim Hughes, Assisted by ♪ Melody ♫
Collection of ♪ Melody ♫, Applewood, CO, USA

QUARTZ - (Included) - [BLACK PHANTOM] - SiO$_2$ with undefined black inclusions) - Vitreous; Hardness 7; Location: Minas Gerais, Brasil

Photography by Jim Hughes, Assisted by ♪ Melody ♫ Collection of ♪ Melody ♫, Applewood, CO, USA

QUARTZ - (Included) [CHLORITE PHANTOM] - SiO$_2$ with Chlorite inclusions) - Vitreous; Hardness 7; Location: Norway

Photography by Jim Hughes, Assisted by ♪ Melody ♫ Collection of ♪ Melody ♫, Applewood, CO, USA

QUARTZ - (Included) [CHLORITE PHANTOM] - SiO$_2$ with Chlorite inclusions) - Vitreous; Hardness 7; Location: Madagascar

Photography by Jim Hughes, Assisted by ♪ Melody ♫ Collection of ♪ Melody ♫, Applewood, CO, USA Gift of Monika Friedrich, RSA & Madagascar

QUARTZ - (Included) - [RED PHANTOM] - SiO_2 with Limonite after Hematite via Kaolinite tubes - Vitreous; Hardness 7; Location: Minas Gerais, Brasil

Photography by Jim Hughes, Assisted by ♪ Melody ♫
Collection of ♪ Melody ♫, Applewood, CO, USA

QUARTZ - (Included) - [RED PHANTOM] - SiO_2 with Limonite after Hematite via Kaolinite tubes - Vitreous; Hardness 7; Location: Minas Gerais, Brasil

Photography by Jim Hughes, Assisted by ♪ Melody ♫
Collection of ♪ Melody ♫, Applewood, CO, USA

QUARTZ - (Included) [PHANTOMS] - SiO_2 with various undefined inclusions - Vitreous; Hardness 7; Location: Minas Gerais, Brasil, Madagascar

Photography by Franklin King, Assisted by Monika Friedrich, Madagascar
Collection of Franklin King & Monika Friedrich

QUARTZ - (Included) [RUTILATED] - SiO₂ with TiO₂ and sometimes FeTiO₃ (acicular yellow/red/white rutile crystals inclusion); Hardness 7; Location: Minas Gerais, Brasil

QUARTZ - (Included) [RUTILATED] - SiO₂ with TiO₂ and sometimes FeTiO₃ (acicular yellow/red/white rutile crystals inclusion); Hardness 7; Location: Minas Gerais, Brasil

QUARTZ - (Included) [RUTILATED] - SiO₂ with TiO₂ and sometimes FeTiO₃ (acicular yellow rutile crystals inclusion); Hardness 7; Location: Minas Gerais, Brasil

QUARTZ - (Included) [RUTILATED] - SiO_2 with TiO_2 and sometimes $FeTiO_3$ (acicular yellow rutile crystals inclusion); Hardness 7; Location: Minas Gerais, Brasil

Photography by Jim Hughes, Assisted by ♪ Melody ♫
Collection of ♪ Melody ♫, Applewood, CO, USA

QUARTZ - (Included) [RUTILATED SMOKEY] - SiO_2 with organic inclusions or due to natural exposure to radioactivity, with TiO_2 as acicular yellow rutile crystals; Hardness 7; Location: Minas Gerais, Brasil

Photography by Jim Hughes, Assisted by ♪ Melody ♫
Collection of ♪ Melody ♫, Applewood, CO, USA

QUARTZ - (Included) [SNOWBALL QUARTZ with Orange Phantom] - SiO_2 with organic inclusions; Hardness 7; Location: Minas Gerais, Brasil

Photography by Jim Hughes, Assisted by ♪ Melody ♫
Collection of ♪ Melody ♫, Applewood, CO, USA

QUARTZ - (Included) [SNOWBALL QUARTZ] - SiO_2 with white organic inclusions; Hardness 7; Location: Arkansas, USA

Photography by Jim Hughes, Assisted by ♪ Melody ♫
Collection of ♪ Melody ♫, Applewood, CO, USA

QUARTZ - (Included) [AMETHYST STRIPED QUARTZ] - SiO_2 with trace amounts of Ferric Iron; Hardness 7; Location: Minas Gerais, Brasil

Photography by Jim Hughes, Assisted by ♪ Melody ♫
Collection of ♪ Melody ♫, Applewood, CO, USA

QUARTZ - (Included) [TOURMALINATED QUARTZ] - SiO_2 with tourmaline inclusions; Hardness 7; Location: Minas Gerais, Brasil

Photography by Jim Hughes, Assisted by ♪ Melody ♫
Collection of ♪ Melody ♫, Applewood, CO, USA

QUARTZ - (Included)
[TOURMALINATED QUARTZ] -
SiO_2 with tourmaline inclusions;
Hardness 7; Location: Minas
Gerais, Brasil

Photography by Jim Hughes, Assisted by ♪ Melody ♫
Collection of ♪ Melody ♫, Applewood, CO, USA

QUARTZ - (WINDOW QUARTZ
CRYSTAL) (Smokey Quartz);
SiO_2 with organic impurities or
due to natural exposure to
radioactivity; Hardness 7;
Location: Montana, USA

Photography by Jim Hughes, Assisted by ♪ Melody ♫
Collection of ♪ Melody ♫, Applewood, CO, USA

QUARTZITE - SiO_2 - Finely
compacted grains with possible
impurities providing opacity to
translucence; Hardness 7;
Locality: Idaho, USA

Photography by Jim Hughes, Assisted by ♪ Melody ♫
Collection of ♪ Melody ♫, Applewood, CO, USA

If it is to be, it's up to me.

[Rev. Sally David, Florida, USA]

RALSTONITE - $Na_{0.4}(Al,Mg)_2$ $(F,OH)_8 \cdot H_2O$ - *Vitreous colourless, white; Hardness 4.5; Locality: Oslo, Norway*

Photography by Jim Hughes, Assisted by ♪ Melody ♫
Collection of ♪ Melody ♫, Applewood, CO, USA

RAMSDELLITE - *gamma-MnO_2 - Grey, black; Hardness 2-4; Locality: New Mexico, USA*

Photography by Jim Hughes, Assisted by ♪ Melody ♫
Collection of ♪ Melody ♫, Applewood, CO, USA

RASPITE - $PbWO_4$ - *Adamantine <u>rust</u>, yellowish-brown, light yellow, grey; Hardness 2.5-3; Locality: New South Wales, Australia*

Photography by Jim Hughes, Assisted by ♪ Melody ♫
Collection of ♪ Melody ♫, Applewood, CO, USA

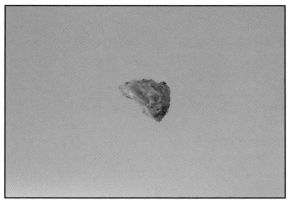

REDDINGITE - $Mn_3(PO_4)_2 \cdot 3H_2O$ *(In Phosphorferrite)* - *Vitreous/sub-resinous pink, yellowish-white to grey, colourless; Hardness 3-3.5; Locality: Bavaria, Germany*

Photography by Jim Hughes, Assisted by ♪ Melody ♫
Collection of Colorado School Of Mines, CO, USA

RHODOCHROSITE - *(Slice)* $MnCo_3$ - *Vitreous pink, rose, red, yellowish-grey, brown; Hardness 3.5-4; Locality: Uruguay, South America*

Photography by Jim Hughes, Assisted by ♪ Melody ♫
Collection of ♪ Melody ♫, Applewood, CO, USA

RHODOCHROSITE - *(With Matrix)* - $MnCo_3$ - *Vitreous pink, rose, red, yellowish-grey, brown; Hardness 3.5-4; Locality: Colorado, USA*

Photography by Jim Hughes, Assisted by ♪ Melody ♫
Collection of Bob Jackson, Applewood, CO, USA

RHODOCHROSITE -
(Stalactite) - MnCo₃ - Vitreous pink, rose, red, yellowish-grey, brown; Hardness 3.5-4; Locality: Uruguay, South America

Photography by Jim Hughes, Assisted by ♪ Melody ♫
Collection of ♪ Melody ♫, Applewood, CO, USA

RHODOCHROSITE - *(With Quartz) - MnCo₃ - Vitreous pink, rose, red, yellowish-grey, brown; Hardness 3.5-4; Locality: Colorado, USA*

Photography by Jim Hughes, Assisted by ♪ Melody ♫
Collection of ♪ Melody ♫, Applewood, CO, USA

RHODOCHROSITE - *(Crystal) - MnCo₃ - Vitreous pink, rose, red, yellowish-grey, brown; Hardness 3.5-4; Locality: Zimbabwe, Africa*

Photography by Jim Hughes, Assisted by ♪ Melody ♫
Collection of ♪ Melody ♫, Applewood, CO, USA

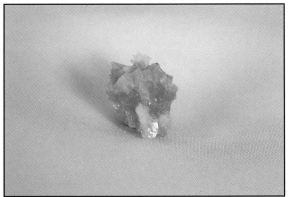

RHODOCHROSITE - *(Cluster)* - *MnCo₃ - Vitreous pink, rose, red, yellowish-grey, brown; Hardness 3.5-4; Locality: Colorado, USA*

Photography by Jim Hughes, Assisted by ♪ Melody ♫
Collection of Bob Jackson, Applewood, CO, USA

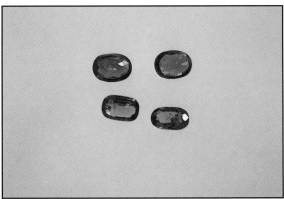

RHODOLITE - *(Faceted)*
$2[3MgO \heartsuit Al_2O_3 \heartsuit 3SiO_2] + [3FeO \heartsuit Al_2O_3 \heartsuit 3SiO_2]$ - *Vitreous to resinous pale rose-red and purple; Hardness 6.5-7.5; Locality: India*

Photography by Jim Hughes, Assisted by ♪ Melody ♫
Collection of Bob Jackson, Applewood, CO, USA

RHODOLITE -
$2[3MgO \heartsuit Al_2O_3 \heartsuit 3SiO_2] + [3FeO \heartsuit Al_2O_3 \heartsuit 3SiO_2]$ - *Vitreous to resinous pale rose-red and purple; Hardness 6.5-7.5; Locality: Mexico*

Photography by Jim Hughes, Assisted by ♪ Melody ♫
Collection of ♪ Melody ♫, Applewood, CO, USA
Gift of LaSonda Sioux Sipe, Virginia, USA

RHODONITE -
$(Mn,Fe,Mg,Ca)SiO_3$ - *Vitreous rose-pink, brownish-red; Hardness 5.5-6.5; Locality: Mexico*

Photography by Jim Hughes, Assisted by ♪ Melody ♫
Collection of ♪ Melody ♫, Applewood, CO, USA

RHYOLITE - *Volcanic fine-grained igneous rock consisting chiefly of alkaline feldspars and quartz. Locality: Idaho, USA*

Photography by Jim Hughes, Assisted by ♪ Melody ♫
Collection of Bob Jackson, Applewood, CO, USA

RICHTERITE -
$Na_2Ca(Mg,Fe)_5Si_8O_{22}(OH)_2$ - *Translucent brown, yellow, brownish-red, green; Hardness 5-6; Locality: Republic of South Africa, Africa*

Photography by Jim Hughes, Assisted by ♪ Melody ♫
Collection of ♪ Melody ♫, Applewood, CO, USA

RICHTERITE -
$Na_2Ca(Mg,Fe)_5Si_8O_{22}(OH)_2$ -
*Translucent brown, yellow,
brownish-red, green; Hardness 5-
6; Locality: Republic of South
Africa, Africa*

Photography by Jim Hughes, Assisted by ♪ Melody ♫
Collection of Bob Jackson, Applewood, CO, USA

RICKARDITE - Cu_7Te_5 - *Metallic
purple-red; Hardness 3.5;
Locality: Colorado, USA*

Photography by Jim Hughes, Assisted by ♪ Melody ♫
Collection of Colorado School of Mines, CO, USA

RIEBECKITE -
$Na_2(Fe^{2+},Mg)_3Fe_2^{3+}Si_8O_{22}(OH)_2$ -
*Vitreous blue; Hardness 5;
Locality: Quebec, Canada*

Photography by Jim Hughes, Assisted by ♪ Melody ♫
Collection of ♪ Melody ♫, Applewood, CO, USA

ROSASITE -$(Cu,Zn)_2CO_3(OH)_2$ - *Green, bluish-green, blue; Hardness 4.5; Locality: Mexico*

Photography by Jim Hughes, Assisted by ♪ Melody ♫
Collection of ♪ Melody ♫, Applewood, CO, USA

ROSCHERITE *(On Rose and Clear Quartz)* - $Ca(Mg,Fe)_2Be_2Al_x(PO_4)_3(OH)_3$♥ $2H_2O$ - *Translucent dark brown, olive-green; Hardness 4.5; Locality: Minas Gerais, Brasil*

Photography by Jim Hughes, Assisted by ♪ Melody ♫
Collection of ♪ Melody ♫, Applewood, CO, USA

ROSE QUARTZ - SiO_2 *with Manganese or Titanium; Light rose to deep rose colour; Hardness 7; Locality: Minas Gerais, Brasil*

Photography by Jim Hughes, Assisted by ♪ Melody ♫
Collection of ♪ Melody ♫, Applewood, CO, USA

ROSE QUARTZ - *(Cluster) SiO$_2$ with Manganese or Titanium; Light rose to deep rose colour; Hardness 7; Locality: Bahia, Brasil*

Photography by Jim Hughes, Assisted by ♪ Melody ♫
Collection of ♪ Melody ♫, Applewood, CO, USA

ROSE QUARTZ - *(With Smokey Quartz) SiO$_2$ with Manganese or Titanium and Smokey Quartz; Rose quartz is light rose to deep rose colour; Hardness 7; Locality: Bahia, Brasil*

Photography by Jim Hughes, Assisted by ♪ Melody ♫
Collection of Jackson & Melody, Applewood, CO, USA

ROSE QUARTZ - *(Buttons) SiO$_2$ with Manganese or Titanium; Light rose to deep rose colour; Hardness 7; Locality: Bahia, Brasil*

Photography by Jim Hughes, Assisted by ♪ Melody ♫
Collection of Bob Jackson, Applewood, CO, USA
Gift of LaSonda Sioux Sipe, Virginia USA

ROSE QUARTZ - SiO₂ with Manganese or Titanium; (With Amethyst, Smokey Quartz, Clear Quartz): Hardness 7; Locality: Bahia, Brasil
Photography by Jim Hughes, Assisted by ♪ Melody ♫
Collection of Jackson & Melody, Applewood, CO, USA

ROSE QUARTZ (ELESTIAL) - SiO₂ with Manganese or Titanium and organic impurities; Locality: Bahia, Brasil
Photography by Jim Hughes, Assisted by ♪ Melody ♫
Collection of ♪ Melody ♫, Applewood, CO, USA

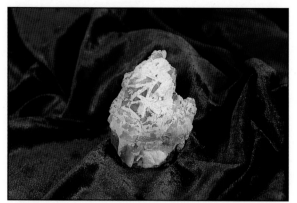

ROSE QUARTZ (ELESTIAL) -
SiO$_2$ with Manganese or Titanium and organic impurities; Hardness 7; Locality: Bahia, Brasil

Photography by Jim Hughes, Assisted by ♪ Melody ♫
Collection of ♪ Melody ♫, Applewood, CO, USA

ROSELITE -
Ca$_2$(Co,Mg)(AsO$_4$)$_2$♥2H$_2$O - Vitreous pink; Hardness 3.5; Locality: Morocco

Photography by Jim Hughes, Assisted by ♪ Melody ♫
Collection of ♪ Melody ♫, Applewood, CO, USA

ROSENBUSCHITE -
(Ca,Na)$_6$TiZr(Si$_2$O$_7$)$_2$(F,OH)$_4$ - Vitreous light orange-grey; Hardness 5-6; Locality: Norway

Photography by Jim Hughes, Assisted by ♪ Melody ♫
Collection of ♪ Melody ♫, Applewood, CO, USA

RUBY - Al₂O₃ - Transparent to translucent red to ruby colour; Hardness 9; Locality: Sri Lanka

Photography by Jim Hughes, Assisted by ♪ Melody ♫
Collection of Bob Jackson, Applewood, CO, USA

RUBY (Crystal) - Al₂O₃ - Translucent red to ruby colour; Hardness 9; Locality: Merlin's Crystal Mine, India

Photography by Jim Hughes, Assisted by ♪ Melody ♫
Collection of ♪ Melody ♫, Applewood, CO, USA

RUBY IN KYANITE - Al₂O₃ with Al₂SiO₅; Hardness 5-9; Locality: India

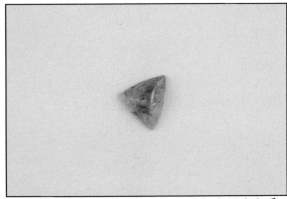

Photography by Jim Hughes, Assisted by ♪ Melody ♫
Collection of ♪ Melody ♫, Applewood, CO, USA
Gift of Ms Sigari, Germany

RUBY SILVER (Also Known As Proustite) - Ag₃AsS₃ - Adamantine translucent red when lit, grey metallic in normal lighting; Hardness 2-2.5; Locality: Mexico; Gift from Angel Torrecillas, Mexico; Photography by Jim Hughes, Assisted by ♪ Melody ♫; Collection of ♪ Melody ♫, Applewood, CO, USA

RUTILE - TiO$_2$ and sometimes with FeTiO$_3$ - Metallic/adamantine red, brown, yellowish, black, etc.; Hardness 6-6.5; Locality: Georgia, USA

Photography by Jim Hughes, Assisted by ♪ Melody ♫
Collection of Bob Jackson, Applewood, CO, USA

RUTILE - TiO$_2$ and sometimes with FeTiO$_3$ - Metallic/adamantine red, brown, yellowish, black, etc.; Hardness 6-6.5; Locality: Georgia, USA

Photography by Jim Hughes, Assisted by ♪ Melody ♫
Collection of ♪ Melody ♫, Applewood, CO, USA

SAGENITE - *SiO₂ with reticulated groups of slender crystals or vacancies of same; Hardness 6-6.5 Locality: California, USA*

Photography by Jim Hughes, Assisted by ♪ Melody ♫
Collection of Bob Jackson, Applewood, CO, USA

SAGENITE - *(Plume) - SiO₂ with reticulated groups of slender crystals or vacancies of same; Hardness 6-6.5 Locality: California, USA*

Photography by Jim Hughes, Assisted by ♪ Melody ♫
Collection of Bob Jackson, Applewood, CO, USA

SAINFELDITE - *(With Guerinite and Rauenthalite Crystals) - Ca₅(AsO₄)₂(AsO₃OH)₂♥4H₂O - Transparent colourless, light pink; Hardness ?; Locality: Alsace, France*

Photography by Jim Hughes, Assisted by ♪ Melody ♫
Collection of ♪ Melody ♫, Applewood, CO, USA

SALESITE - *(Crystals);* $CuIO_3(OH)$ - *Transparent bluish-green; Hardness 3; Locality: Chile, South America*

Photography by Jim Hughes, Assisted by ♪ Melody ♫
Collection of ♪ Melody ♫, Applewood, CO, USA

SAMARSKITE - $(Y,Ce,U,Fe)_3(Nb,Ta,Ti)_5O_{16}$ - *Vitreous/resinous black; Hardness 5-6; Locality: Maine, USA*

Photography by Jim Hughes, Assisted by ♪ Melody ♫
Collection of ♪ Melody ♫, Applewood, CO, USA

SANDSTONE - *Sand grains bound together with silica, carbonate, etc.; Locality: Mexico*

Photography by Jim Hughes, Assisted by ♪ Melody ♫
Collection of Bob Jackson, Applewood, CO, USA

SAPPHIRE - Al_2O_3 - <u>(Blue-Yellow PARTI)</u>; *Hardness 9; Locality: New South Wales, Australia*

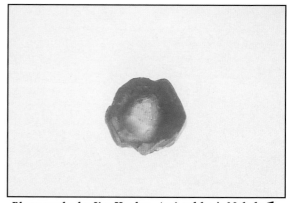

Photography by Jim Hughes, Assisted by ♪ Melody ♫
Collection of ♪ Melody ♫, Applewood, CO, USA

SAPPHIRE - Al_2O_3 - <u>(Blue-Yellow-White PARTI)</u>; *Hardness 9; Locality: New South Wales, Australia*

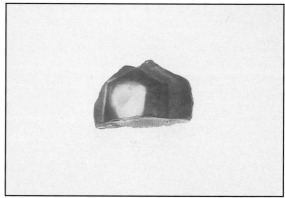

Photography by Jim Hughes, Assisted by ♪ Melody ♫
Collection of ♪ Melody ♫, Applewood, CO, USA

SAPPHIRE - Al_2O_3 - <u>(Blue-Yellow-White-Gold-Black PARTI)</u>; *Hardness 9; Locality: New South Wales, Australia*

Photography by Jim Hughes, Assisted by ♪ Melody ♫
Collection of ♪ Melody ♫, Applewood, CO, USA

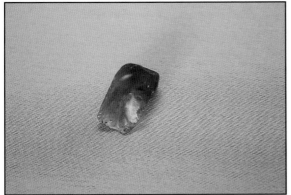

SAPPHIRE - Al_2O_3 - (<u>Green-Yellow PARTI</u>); *Hardness 9; Locality: New South Wales, Australia*

Photography by Jim Hughes, Assisted by ♪ Melody ♫
Collection of ♪ Melody ♫, Applewood, CO, USA

SAPPHIRE - Al_2O_3 - (<u>Green PARTI</u>); *Hardness 9; Locality: New South Wales, Australia*

Photography by Jim Hughes, Assisted by ♪ Melody ♫
Collection of ♪ Melody ♫, Applewood, CO, USA

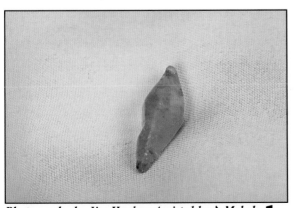

SAPPHIRE - Al_2O_3 - (<u>Blue-Green</u>); *Hardness 9; Locality: Sri Lanka*

Photography by Jim Hughes, Assisted by ♪ Melody ♫
Collection of ♪ Melody ♫, Applewood, CO, USA

SAPPHIRE - Al_2O_3 - (<u>Deep-Blue</u>); Hardness 9; Locality: Sri Lanka

Photography by Jim Hughes, Assisted by ♪ Melody ♫
Collection of ♪ Melody ♫, Applewood, CO, USA

SAPPHIRE - Al_2O_3 - (<u>Blue-Star</u>); Hardness 9; Locality: Burma

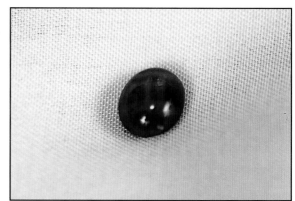

Photography by Jim Hughes, Assisted by ♪ Melody ♫
Collection of Bob Jackson, Applewood, CO, USA

SAPPHIRE - Al_2O_3 - (<u>Black</u>); Hardness 9; Locality: India

Photography by Jim Hughes, Assisted by ♪ Melody ♫
Collection of Bob Jackson, Applewood, CO, USA

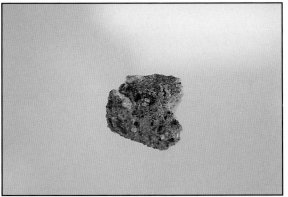

SARCOLITE -
$(Ca,Na)_8Al_4Si_6O_{23}$
$(PO_4,CO_3,SO_4,F,Cl,OH,H_2O)_2$ -
Vitreous pink, red, reddish-white;
Hardness 6; Locality: Italy

Photography by Jim Hughes, Assisted by ♪ Melody ♫
Collection of Colorado School of Mines, CO, USA

SARDONYX - SiO_2 **-** *Even planed*
banded straight transparent to
translucent white, brown, black,
with layers of carnelian; Hardness
of 7; Locality, Minas Gerais,
Brasil

Photography by Jim Hughes, Assisted by ♪ Melody ♫
Collection of ♪ Melody ♫, Applewood, CO, USA

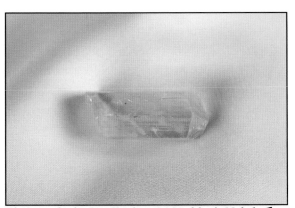

SCAPOLITE -
$(Na,Ca)_4(Si,Al)_{12}O_{24}(Cl,CO_3,SO_4)$ -
[General silicate with additional
anion] Colour range includes
pink, golden, yellow, white,
colourless, green, violet, yellow,
grey, blue, red, etc.; Hardness
from 5-6.5; Locality: Tanzania,
Africa

Photography by Jim Hughes, Assisted by ♪ Melody ♫
Collection of ♪ Melody ♫, Applewood, CO, USA
Gift of David Pearl, Kenya, Africa

SCAPOLITE -
(Na,Ca)$_4$(Si,Al)$_{12}$O$_{24}$(Cl,CO$_3$,SO$_4$)
- [General silicate with additional anion] Colour range includes pink, golden, yellow, white, colourless, green, violet, yellow, grey, blue, red, etc.; Hardness 5-6.5; Locality: Madagascar

Photography by Jim Hughes, Assisted by ♪ Melody ♫
Collection of ♪ Melody ♫, Applewood, CO, USA

SCAWTITE -
Ca$_7$(Si$_3$O$_9$)$_2$(CO$_3$)♥2H$_2$O -
Vitreous colourless; Hardness 4.5-5; Locality: California, USA

Photography by Jim Hughes, Assisted by ♪ Melody ♫
Collection of ♪ Melody ♫, Applewood, CO, USA

SCHALENBLENDE -
Combination of Galena and Jasper - Hardness varies from 2.6 to 7; Locality: Germany

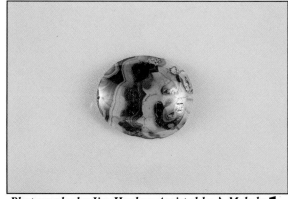

Photography by Jim Hughes, Assisted by ♪ Melody ♫
Collection of ♪ Melody ♫, Applewood, CO, USA
Gift of Ms Sigari, Germany

SCHALENBLENDE -
Combination of Galena and
Jasper - Hardness varies from 2.6
to 7; Locality: Germany

Photography by Jim Hughes, Assisted by ♩ Melody ♫
Collection of Bob Jackson, Applewood, CO, USA

SCHEELITE - CaWO₄ -
Vitreous/adamantine colourless,
white, yellow, greenish, etc.;
Hardness 4.5-5; Locality: Korea

Photography by Jim Hughes, Assisted by ♩ Melody ♫
Collection of ♩ Melody ♫, Applewood, CO, USA

SCHIST - A form of Mica,
probably Margarodite; Hardness
estimated to be 2-2.5; Locality:
Idaho, USA

Photography by Jim Hughes, Assisted by ♩ Melody ♫
Collection of ♩ Melody ♫, Applewood, CO, USA

SEAMANITE -
$Mn_3[B(OH)_4](PO_4)(OH)_2$ -
Transparent pale yellow, wine-yellow; Hardness 4; Locality: Michigan, USA

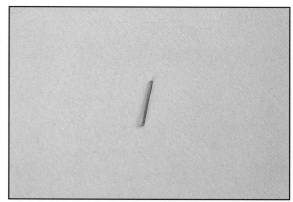

Photography by Jim Hughes, Assisted by ♪ Melody ♫
Collection of ♪ Melody ♫, Applewood, CO, USA

SELENITE - (Cluster) -
$CaSO_4 \heartsuit 2H_2O$- *Crystallized - Transparent colourless with some flexibility; Hardness 1.5-2; Locality: Mexico*

Photography by Jim Hughes, Assisted by ♪ Melody ♫
Collection of ♪ Melody ♫, Applewood, CO, USA

SELENITE - (Curved Crystal) -
$CaSO_4 \heartsuit 2H_2O$ *crystallized - Transparent colourless with some flexibility; Hardness 1.5-2; Locality: Mexico*

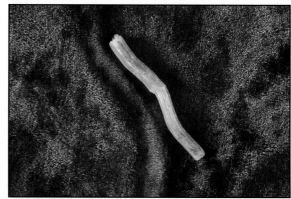

Photography by Jim Hughes, Assisted by ♪ Melody ♫
Collection of ♪ Melody ♫, Applewood, CO, USA

SELENITE - (Cross with undefined druse) - CaSO₄♥2H₂O crystallized - Transparent colourless with some flexibility; Hardness 1.5-2; Locality: Mexico

SELENITE - (Cross with undefined druse) - $CaSO_4 \cdot 2H_2O$ crystallized - Transparent colourless with some flexibility; Hardness 1.5-2; Locality: Mexico

*Photography by Jim Hughes, Assisted by ♪ Melody ♫
Collection of ♪ Melody ♫, Applewood, CO, USA*

SELENITE - (Nodule Cluster) - $CaSO_4 \cdot 2H_2O$ crystallized - Transparent colourless with some flexibility; Hardness 1.5-2; Locality: Minas Gerais, Brasil

*Photography by Jim Hughes, Assisted by ♪ Melody ♫
Collection of ♪ Melody ♫, Applewood, CO, USA*

SELLAITE - MgF_2 - Vitreous colourless, white to yellow; Hardness 5; Locality: Bahia, Brasil

*Photography by Jim Hughes, Assisted by ♪ Melody ♫
Collection of ♪ Melody ♫, Applewood, CO, USA*

SENARMONITE - Sb_2O_3 -
Resinous colourless, greyish-white; Hardness 2-2.5; Locality: Algeria, Africa

SERENDIBITE - *(With Phlogopite)* -
$Ca_2(Mg,Al)_6(Si,Al,B)_6O_{20}$ -
Translucent blue; Hardness 6.7; Locality: New York, USA

SERANDITE - *Hydrous Silicate of Manganese, Calcium, and Sodium - Rose-red in colour; Hardness 3-4; Locality: Quebec, Canada*

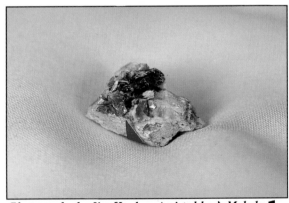

SERPENTINE - *(With Chrysotite)* $(Mg,Al,Fe,Mn,Ni,Zn)_{2-3}$ $(Si,Al,Fe)_2O_5(OH)_4$ - *[General sheet silicate]; Colour range includes green, blackish-green, brownish-red, brownish-yellow, white, yellowish-grey; Hardness 2.5-4; Locality: Vermont, USA*

Photography by Jim Hughes, Assisted by ♪ Melody ♫
Collection of ♪ Melody ♫, Applewood, CO, USA

SERPENTINE - $(Mg,Al,Fe,Mn,Ni,Zn)_{2-3}$ $(Si,Al,Fe)_2O_5(OH)_4$ - *[General sheet silicate]; Colour range includes green, blackish-green, brownish-red, brownish-yellow, white, yellowish-grey; Hardness 2.5-4; Locality: Oregon, USA*

Photography by Jim Hughes, Assisted by ♪ Melody ♫
Collection of ♪ Melody ♫, Applewood, CO, USA
Gift from Bob Jackson, Applewood, CO, USA

SHATTUCKITE - *(With Ajoite [green] and Tenorite [black]);* $Cu_5(SiO_3)_4(OH)_2$ - *Translucent dark blue; Hardness 3.5; Locality: Bisbee, AZ*

Photography by Jim Hughes, Assisted by ♪ Melody ♫
Collection of ♪ Melody ♫, Applewood, CO, USA
Gift from Bob Jackson, Applewood, CO, USA

SHATTUCKITE -
$Cu_5(SiO_3)_4(OH)_2$ - *Translucent dark blue; Hardness 3.5; Locality: Bisbee, AZ*

Photography by Jim Hughes, Assisted by ♪ Melody ♫
Collection of ♪ Melody ♫, Applewood, CO, USA

SIDERITE - $FeCO_3$ **-** *Vitreous yellowish-brown, grey, pale green, white; Hardness 4; Locality: Colorado, USA*

Photography by Jim Hughes, Assisted by ♪ Melody ♫
Collection of ♪ Melody ♫, Applewood, CO, USA

SIDERITE - *(Included In Faceted Quartz)* **-** $FeCO_3$ **-** *Vitreous yellowish-brown, grey, pale green, white; Hardness 4; Locality, Minas Gerais, Brasil*

Photography by Jim Hughes, Assisted by ♪ Melody ♫
Collection of ♪ Melody ♫, Applewood, CO, USA

SILLIMANITE - *(With Garnet)* - *Al_2SiO_5 - Translucent colourless, white, yellow, brown, green; Hardness 6.5-7.5; Locality: Minas Gerais, Brasil*

Photography by Jim Hughes, Assisted by ♪ Melody ♫
Collection of ♪ Melody ♫, Applewood, CO, USA

SILVER *(Nugget) - Ag - Metallic white; Hardness 2.5-3; Locality: Michigan, USA*

Photography by Jim Hughes, Assisted by ♪ Melody ♫
Collection of ♪ Melody ♫, Applewood, CO, USA

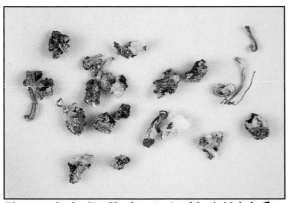

SILVER *(Native) - Ag - Metallic white; Hardness 2.5-3; Locality: Mexico*

Photography by Jim Hughes, Assisted by ♪ Melody ♫
Collection of ♪ Melody ♫, Applewood, CO, USA
Gift of Angel Torrecillas, Mexico

SILVER - *(With Quartz and Cobalt) - Ag - Metallic white; Hardness 2.5-3; Locality: Nevada, USA*

Photography by Jim Hughes, Assisted by ♪ Melody ♫
Collection of Bob Jackson, Applewood, CO, USA

SILVER - *(Plume In Quartzite) - Ag - Metallic white; Hardness from 2.5-3; Locality: Nevada, USA*

Photography by Jim Hughes, Assisted by ♪ Melody ♫
Collection of Bob Jackson, Applewood, CO, USA

SILVER - *(In Quartz Crystal) - Ag - Metallic white; Hardness from 2.5-3; Locality: Western USA*

Photography by Jim Hughes, Assisted by ♪ Melody ♫
Collection of ♪ Melody ♫, Applewood, CO, USA

SIMPSONITE -
$Al_4(Ta,Nb)_3O_{13}(F,OH)$ -
Transparent colourless, cream;
Hardness ?; Locality: Rio Grande
Do Norte, Brasil

Photography by Jim Hughes, Assisted by ♪ Melody ♫
Collection of ♪ Melody ♫, Applewood, CO, USA

SKAN - Composition not known
to author; Locality: Siberia

Photography by Jim Hughes, Assisted by ♪ Melody ♫
Collection of ♪ Melody ♫, Applewood, CO, USA
Gift from Ms Sigari, Germany

SMITHSONITE - $ZnCO_3$ -
Vitreous greyish-white, <u>green-</u>
<u>blue</u>, *brown, pink, etc.; Hardness*
4-4.5, Locality, Mexico

Photography by Jim Hughes, Assisted by ♪ Melody ♫
Collection of ♪ Melody ♫, Applewood, CO, USA

SMITHSONITE - $ZnCO_3$ - *Vitreous greyish-white, green-blue, brown, <u>pink</u>, etc.; Hardness 4-4.5; Locality: Namibia, Africa*

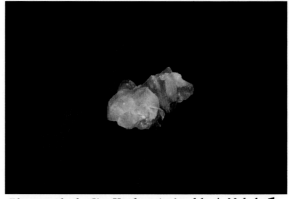

Photography by Jim Hughes, Assisted by ♪ Melody ♫
Collection of Bob Jackson, Applewood, CO, USA

SMOKEY QUARTZ - *(Crystals) -SiO_2 with organic impurities or due to natural exposure to radioactivity; Transparent to opaque smokey yellow to dark smokey brown to brownish-black to black; Hardness 7; Locality: Sinai, Middle East*

Photography by Jim Hughes, Assisted by ♪ Melody ♫
Collection of ♪ Melody ♫, Applewood, CO, USA

SMOKEY QUARTZ - *(Crystal) (With Topaz) - SiO_2 with organic impurities or due to natural exposure to radioactivity; Hardness 7; Locality: Namibia, Africa*

Photography by Jim Hughes, Assisted by ♪ Melody ♫
Collection of ♪ Melody ♫, Applewood, CO, USA

SMOKEY QUARTZ - *(Crystal)* - *(With Citrine; Crystal form)* - *SiO₂ with organic impurities or due to natural exposure to radioactivity and with colloidal iron hydrates ; Hardness 7; Locality; Minas Gerais, Brasil*

Photography by Jim Hughes, Assisted by ♪ Melody ♫
Collection of ♪ Melody ♫, Applewood, CO, USA

SMOKEY QUARTZ - *(Tabby Crystal)* - *SiO₂ with organic impurities or due to natural exposure to radioactivity; Transparent to opaque smokey yellow to dark smokey brown to brownish-black to black; Hardness 7; Locality: Zaire, Africa*

Photography by Jim Hughes, Assisted by ♪ Melody ♫
Collection of ♪ Melody ♫, Applewood, CO, USA

SMOKEY/CLEAR QUARTZ - *(Faceted)* - *SiO₂ with organic impurities or due to natural exposure to radioactivity with pure SiO₂ ; Hardness 7; Locality: Minas Gerais, Brasil*

Photography by Jim Hughes, Assisted by ♪ Melody ♫
Collection of ♪ Melody ♫, Applewood, CO, USA
Gift of Ricardo Foscarini de Almeida, Teofilo Otoni, Minas Gerais, Brasil

SMOKEY/CLEAR QUARTZ - SiO_2 *with organic impurities or due to natural*
exposure to radioactivity with pure SiO_2 *- Hardness 7; Locality: Minas Gerais, Brasil*
Photography by Jim Hughes, Assisted by ♪ Melody ♫
Collection of ♪ Melody ♫, Applewood, CO, USA
Gift of Jose Orizon de Almeida, Teofilo Otoni, Minas Gerais, Brasil

SODALITE - $Na_4(Si_3Al_3O_{12}Cl$ - Translucent blue, pink, grey, yellow, green, white; Hardness 5.5-6; Locality; Minas Gerais, Brasil

Photography by Jim Hughes, Assisted by ♪ Melody ♫
Collection of Bob Jackson, Applewood, CO, USA

SONOLITE - $Mn_9(SiO_4)_4(OH,F)_2$ - (With Franklinite and Zincite); Translucent pinkish-brown; Hardness 5.5; Locality: New Jersey, USA

Photography by Jim Hughes, Assisted by ♪ Melody ♫
Collection of ♪ Melody ♫, Applewood, CO, USA

SPANGOLITE - (In Quartz) $Cu_6AlSO_4(OH)_{12}Cl♥3H_2O$ - Vitreous dark green, bluish-green; Hardness 2; Locality: New Mexico, USA

Photography by Jim Hughes, Assisted by ♪ Melody ♫
Collection of ♪ Melody ♫, Applewood, CO, USA

SPESSARTINE/SPESSARTITE
(Garnet) - Mn₃Al₂(SiO₄)₃ -
Vitreous/resinous dark red,
<u>*brownish-red*</u>*; Hardness 6.5-7.5;*
Locality: Bavaria, Germany

Photography by Jim Hughes, Assisted by ♪ Melody ♫
Collection of ♪ Melody ♫, Applewood, CO, USA

SPESSARTINE/SPESSARTITE
(Garnet) - Mn₃Al₂(SiO₄)₃ -
Vitreous/resinous dark red,
brownish-red; Hardness 6.5-7.5;
Locality: Nevada, USA

Photography by Jim Hughes, Assisted by ♪ Melody ♫
Collection of ♪ Melody ♫, Applewood, CO, USA

SPHAEROCOBALTITE -
CoCO₃ - Vitreous rose-red;
Hardness 4; Locality: Liguria,
Italy

Photography by Jim Hughes, Assisted by ♪ Melody ♫
Collection of ♪ Melody ♫, Applewood, CO, USA

SPHALERITE - *(Crystal) - alpha-ZnS - Resinous/adamantine colourless, black, brown, etc.; Hardness 3.5; Locality: Missouri, USA*

Photography by Jim Hughes, Assisted by ♪ Melody ♫
Collection of ♪ Melody ♫, Applewood, CO, USA

SPHALERITE - *(With Orange Calcite) - alpha-ZnS with Calcite; Hardness 3.5; Locality: Mexico*

Photography by Jim Hughes, Assisted by ♪ Melody ♫
Collection of ♪ Melody ♫, Applewood, CO, USA

SPHENE - $CaTiSiO_5$ - *Adamantine to resinous, transparent to opaque light shaded colours; Hardness 5-5.5; Locality: Minas Gerais, Brasil*

Photography by Jim Hughes, Assisted by ♪ Melody ♫
Collection of ♪ Melody ♫, Applewood, CO, USA

SPINEL - *(Black Record Keeper Crystal)* - $MgAl_2O_4$ - Vitreous; Hardness 7.5-8; Locality: Quebec, Canada

Photography by Jim Hughes, Assisted by ♪ Melody ♫
Collection of ♪ Melody ♫, Applewood, CO, USA

SPINEL - *(Green)* - $MgAl_2O_4$ - Vitreous; Hardness 7.5-8; Locality: India

Photography by Jim Hughes, Assisted by ♪ Melody ♫
Collection of ♪ Melody ♫, Applewood, CO, USA

SPINEL - *(Red)* - $MgAl_2O_4$ - Vitreous; Hardness 7.5-8; Locality: Sri Lanka

Photography by Jim Hughes, Assisted by ♪ Melody ♫
Collection of Bob Jackson, Applewood, CO, USA

SPINEL - *(Purple)* - *MgAl₂O₄* -
Vitreous; Hardness 7.5-8;
Locality: Burma

SPODUMENE - *LiAlSi₂O₆* -
Vitreous colourless, yellow,
greyish-white, lilac, pale green,
etc.; Hardness 6.5-7; Locality:
Afghanistan

STANNITE - *Cu₂FeSnS₄* -
Metallic grey/black, olive-grey in
reflected light; Hardness 4;
Locality: Bolivia, South America

STAUROLITE -
$(Fe,Mg)_4Al_{17}(Si,Al)_8O_{44}(OH)_4$ -
Translucent dark/reddish/yellow-brown, pale yellow; Hardness from 7-7.5; Locality: Maine, USA

Photography by Jim Hughes, Assisted by ♪ Melody ♫
Collection of ♪ Melody ♫, Applewood, CO, USA

STAUROLITE -
$(Fe,Mg)_4Al_{17}(Si,Al)_8O_{44}(OH)_4$ -
Translucent dark/reddish/yellow-brown, pale yellow; Hardness from 7-7.5; Locality: Georgia, USA

Photography by Jim Hughes, Assisted by ♪ Melody ♫
Collection of ♪ Melody ♫, Applewood, CO, USA

STEATITE - $H_2Mg_3(SiO_3)_4$ -
Coarse granular greyish-green to brownish-grey; Hardness from 1-2.5; Locality: Oregon, USA

Photography by Jim Hughes, Assisted by ♪ Melody ♫
Collection of Bob Jackson, Applewood, CO, USA

STEPHANITE - Ag_5SbS_4 - *Metallic black, grey in reflected light; Hardness 2-2.5; Locality: Mexico*

Photography by Jim Hughes, Assisted by ♪ Melody ♫
Collection of ♪ Melody ♫, Applewood, CO, USA

STIBICONITE *(After Stibnite)* - $H_2Sb_2O_5$ - *Usually pale yellow to yellow-white; Amorphous and variable in composition and hardness; Locality: Mexico*

Photography by Jim Hughes, Assisted by ♪ Melody ♫
Collection of ♪ Melody ♫, Applewood, CO, USA

STIBIOTANTALITE - $SbTaO_4$ - *Resinous/adamantine brown, reddish-brown, reddish-yellow, greenish-yellow; Hardness 5.5; Locality: California, USA*

Photography by Jim Hughes, Assisted by ♪ Melody ♫
Collection of Colorado School of Mines, CO, USA

STIBNITE - Sb_2S_3 - *Metallic grey, white in polished sections; Hardness 2; Locality: Japan*

STICHTITE - $Mg_6Cr_2CO_3(OH)_{16}♥4H_2O$ - *Waxy/lustrous lilac, rose-pink; Hardness 1.5-2; Locality: Tasmania, Australia*

STILBITE - $NaCa_4(Si_{27}Al_9)O_{72}♥30H_2O$ - *Transparent to translucent <u>colourless</u>, red, yellow, brown; Hardness 3.5-4; Locality: Oregon, USA*

STILLWELLITE -
$(Ce,La,Ca)BSiO_5$ - Translucent
colourless; Hardness ?; Locality:
Queensland, Australia

STRASHIMIRITE - Composition
unknown to author; Colour green;
Locality: Nevada

STRENGITE - $FePO_4$♥$2H_2O$ -
Vitreous red to orange, violet;
Hardness from 3.5-4.5; Locality:
Arkansas, USA

STROMBOLITE - *(Spurrite) -*
Ca₅(SiO₄)₂(CO₃) - *Translucent*
pale grey, purple; Hardness 5;
Locality: Mexico

Photograph courtesy of Bob Simmons and Kathy Warner,
Heaven & Earth, Vermont, USA. Photograph by David
Benoit, Massachusetts, USA

STROMBOLITE - *(Crystalline)*
(Spurrite) -
Ca₅(SiO₄)₂(CO₃) - *Translucent*
pale grey, purple; Hardness 5;
Locality: Mexico

Photography by Jim Hughes, Assisted by ♪ Melody ♫
Collection of ♪ Melody ♫, Applewood, CO, USA
Gift of Ed Maslovicz, Sanctuary Crystals,
Alsip, Illinois, USA

STRONTIANITE - *SrCO₃*
Vitreous to resinous; Colour
ranges in greens; Hardness 3.5-
4; Locality: Illinois, USA

Photography by Jim Hughes, Assisted by ♪ Melody ♫
Collection of ♪ Melody ♫, Applewood, CO, USA

SUGILITE *(With Manganese) -*
$KNa_2Li_3(Fe,Mn,Al)_2Si_{12}O_{30}$ *-*
Vitreous <u>*rose-red*</u>*, light brownish-*
yellow; Hardness 6-6.5; Locality:
Republic Of South Africa, Africa

Photography by Jim Hughes, Assisted by ♪ Melody ♫
Collection of ♪ Melody ♫, Applewood, CO, USA

SUGILITE *(Crystals) -*
$KNa_2Li_3(Fe,Mn,Al)_2Si_{12}O_{30}$ *-*
Vitreous <u>*rose-red*</u> *(with quartz),*
light brownish-yellow; Hardness
6-6.5; Locality: Republic Of South
Africa,

Photography by Jim Hughes, Assisted by ♪ Melody ♫
Collection of ♪ Melody ♫, Applewood, CO, USA
Gift from Rob Smith, African Gems & Minerals,
Johannesburg, Republic Of South Africa

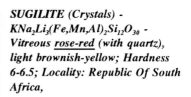

SUGILITE *(Sphere)*
(With Manganese) -
$KNa_2Li_3(Fe,Mn,Al)_2Si_{12}O_{30}$ *-*
Vitreous <u>*rose-red*</u>*, light brownish-*
yellow; Hardness 6-6.5; Locality:
Republic Of South Africa, Africa

Photography by Jim Hughes, Assisted by ♪ Melody ♫
Collection of ♪ Melody ♫, Applewood, CO, USA

SULPHUR - alpha-S - Translucent resinous yellow; Hardness 1.5-2.5;
Locality: Island of Sicily.
Photography by Jim Hughes, Assisted by ♪ Melody ♫
Collection of ♪ Melody ♫, Applewood, CO, USA

SUNSTONE - (Na,Ca)AlSi₃O₈ with inclusions of sub-microscopic lamellae;
Transparent to sub-translucent vitreous and possibly reflective grey, green, <u>yellow</u>,
brown, orange, pink, peach, red; Hardness 5-6; Locality: Canada;
Photography by Jim Hughes, Assisted by ♪ Melody ♫; Collection of ♪ Melody ♫

SUNSTONE - *(Na,Ca)AlSi₃O₈* *with inclusions of sub-microscopic lamellae; Transparent to sub-translucent vitreous and possibly reflective grey, green, <u>yellow,</u> brown, orange, pink, peach, red; Hardness 5-6; Locality: Oregon, USA*

Photography by Jim Hughes, Assisted by ♪ Melody ♫
Collection of ♪ Melody ♫, Applewood, CO, USA
Gift of W.R. Horning, California, USA

SVABITE - *Ca₅(AsO₄)₃F* - *Vitreous/sub-resinous colourless, yellowish-white, grey; Hardness from 4-5; Locality: Langban, Sweden*

Photography by Jim Hughes, Assisted by ♪ Melody ♫
Collection of ♪ Melody ♫, Applewood, CO, USA

A minor deviation in focus...
and one sees beyond illusion to reality ♪

TAAFFEITE - $Mg_3BeAl_8O_{16}$ - Transparent mauve, <u>yellow to yellowish/cream</u>; Hardness from 8 - 8.5; Locality: Peoples Republic Of China

Photography by Jim Hughes, Assisted by ♪ Melody ♫
Collection of ♪ Melody ♫, Applewood, CO, USA

TANTALITE - (Ta); However, the composition $(Fe,Mn)(Nb,Ta)_2O_6$, which represents Columbite, contains the Tantalite in sufficient quantity for function; Sub-metallic to sub-resinous opaque in colour range of brown, red, grey black; Hardness of 6: Locality: Finland

Photography by Jim Hughes, Assisted by ♪ Melody ♫
Collection of ♪ Melody ♫, Applewood, CO, USA

TANZANITE (Crystals) - $Ca_2Al_3Si_3O_{12}(OH)$ with possible impurities - Vitreous blue to purple; Hardness 6-6.5; Locality: Tanzania, Africa

Photography by Jim Hughes, Assisted by ♪ Melody ♫
Collection of ♪ Melody ♫, Applewood, CO, USA

TARBUTTITE - $Zn_2PO_4(OH)$ -
*Vitreous colourless, pale <u>yellow</u>,
brown, blue/grey, red, green;
Hardness 3.7; Locality:
Zimbabwe, Africa*

*Photography by Jim Hughes, Assisted by ♪ Melody ♫
Collection of Bob Jackson, Applewood, CO, USA*

TARBUTTITE - $Zn_2PO_4(OH)$ -
*Vitreous colourless, pale yellow,
<u>blue/grey</u>, brown, red, green;
Hardness 3.7; Locality: Zambia,
Africa*

*Photography by Jim Hughes, Assisted by ♪ Melody ♫
Collection of ♪ Melody ♫, Applewood, CO, USA*

TAVORITE - *(On Altered
Triphylite)* $LiFePO_4(OH)$ -
*Translucent <u>greenish-yellow</u>;
Hardness ?; Locality: South
Dakota, USA*

*Photography by Jim Hughes, Assisted by ♪ Melody ♫
Collection of ♪ Melody ♫, Applewood, CO, USA*

TEKTITE - *Meteoritic glass from outer space; Locality: Pangasinan, Philippines*

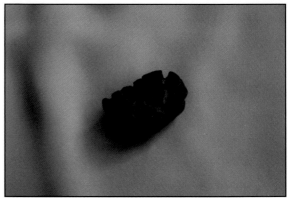

Photography by ♪ Melody ♫
Collection of ♪ Melody ♫, Applewood, CO, USA
Gift from Dan McKee, Philippines

TELLURIUM - *(Native) - Te - Metallic white; Hardness 2-2.5; Locality: Eastern Uzbekistan, Russia*

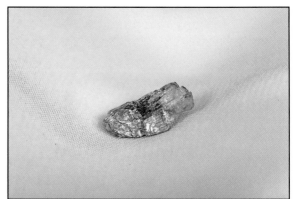

Photography by Jim Hughes, Assisted by ♪ Melody ♫
Collection of ♪ Melody ♫, Applewood, CO, USA

TENNANTITE (On Quartz) - *Cu_3AsS_3 - Isomorphous with Tetrahedrite and possibly Antimony; Hardness 3-4; Locality: Mexico*

Photography by Jim Hughes, Assisted by ♪ Melody ♫
Collection of ♪ Melody ♫, Applewood, CO, USA

TEPHROITE - Mn_2SiO_4 - *Translucent olive-green, bluish-green, green-grey, grey; Hardness 6; Locality: Cornwall, England*

Photography by Jim Hughes, Assisted by ♪ Melody ♫
Collection of ♪ Melody ♫, Applewood, CO, USA

THALENITE - $Y_3Si_3O_{10}(OH)$ - *Translucent flesh-red; Hardness 6.5; Locality: Arizona, USA*

Photography by Jim Hughes, Assisted by ♪ Melody ♫
Collection of ♪ Melody ♫, Applewood, CO, USA

THAUMASITE - $Ca_3Si(OH)_6(CO_3)(SO_4)$♥$12H_2O$ - *Translucent/earthy white to pale yellow; Hardness 3.5; Locality: Republic of South Africa, Africa*

Photography by Jim Hughes, Assisted by ♪ Melody ♫
Collection of Bob Jackson, Applewood, CO, USA

THENARDITE - (After Mirabilite) - alpha-Na₂SO₄ - Vitreous/pearly colourless, white; Hardness 2.7; Locality: Callifornia, USA

Photography by Jim Hughes, Assisted by ♪ Melody ♫
Collection of ♪ Melody ♫, Applewood, CO, USA

THOMSONITE - $NaCa_2(Al_5Si_5)O_{20}♥6H_2O$ - Transparent colourless, peach, yellowish; Hardness 5; Locality: Amudikha Riv. Yakutia Republic, Russia

Photography by Jim Hughes, Assisted by ♪ Melody ♫
Collection of ♪ Melody ♫, Applewood, CO, USA

THOREAULITE - $SnTa_2O_6$ - Resinous/adamantine tan to brown; Hardness 6; Locality: Belgium Congo

Photography by Jim Hughes, Assisted by ♪ Melody ♫
Collection of Colorado School Of Mines, CO, USA

THULITE - $Ca_2Al_3(SiO_4)_3(OH)$ - *Rose coloured, pearly; Hardness 6-6.5; Locality: Norway*

Photography by Jim Hughes, Assisted by ♪ Melody ♫
Collection of Bob Jackson, Applewood, CO, USA

THUNDEREGG - *Fortification agate enclosed by hardened compact matrix usually containing silica ash; Locality: Oregon, USA.*

Photography by Jim Hughes, Assisted by ♪ Melody ♫
Collection of ♪ Melody ♫, Applewood, CO, USA

TIGER EYE - $NaFe^{+3}(SiO_3)_2$ *with impurities - Fibrous, silky to dull luster, colours lavender/blue, green, yellow, red, etc.; Formed as quartz replaces fibers in asbestos; Hardness 4-7; Locality: Griqualand West, Republic Of South Africa*

Photography by Jim Hughes, Assisted by ♪ Melody ♫
Collection of Bob Jackson, Applewood, CO, USA

TIGER EYE - $NaFe^{+3}(SiO_3)_2$ with impurities - Fibrous, silky to dull luster, colours lavender/<u>blue</u>, green, yellow, red, etc.; Formed as quartz replaces fibers in asbestos; Hardness 4-7; Locality: Griqualand West, Republic Of South Africa

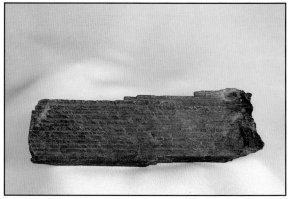

Photography by Jim Hughes, Assisted by ♪ Melody ♫
Collection of Bob Jackson, Applewood, CO, USA

TIGER IRON - Combination of Hematite, Jasper, and Tiger-Eye; Hardness 7; Locality: Australia

Photography by Jim Hughes, Assisted by ♪ Melody ♫
Collection of Bob Jackson, Applewood, CO, USA

TITANITE - $CaTiSiO_5$ - Adamantine to resinous brown, grey, yellow, green rose, red, black; Transparent to opaque; Hardness 5-5.5; Locality: Pakistan

Photography by Jim Hughes, Assisted by ♪ Melody ♫
Collection of ♪ Melody ♫, Applewood, CO, USA

TOPAZ - *(Crystals)* -
$Al_2SiO_4(F,OH)_2$ - *Translucent colourless, white, yellow, grey, green, red, blue; Hardness 8; Locality: Namibia, Africa*

Photography by Jim Hughes, Assisted by ♪ Melody ♫
Collection of ♪ Melody ♫, Applewood, CO, USA

TOPAZ - *(Crystal)* -
$Al_2SiO_4(F,OH)_2$ - *Translucent colourless, white, yellow, grey, green, red, blue; Hardness 8; Locality: Ouro Preto, Minas Gerais, Brasil Belo Horizonte, Minas Gerais, Brasil*

Photography by Jim Hughes, Assisted by ♪ Melody ♫
Collection of ♪ Melody ♫, Applewood, CO, USA
Gift from Gisa, Belo Horizonte, MG, Brasil

TOPAZ - *("Rutilated")* -
$Al_2SiO_4(F,OH)_2$ - *Translucent colourless, white, yellow, grey, green, red, <u>blue</u>; Hardness 8; Locality: Minas Gerais, Brasil*

Photography by Jim Hughes, Assisted by ♪ Melody ♫
Collection of Bob Jackson, Applewood, CO, USA

TOPAZ - ("Rutilated") -
Al₂SiO₄(F,OH)₂ - Translucent
<u>colourless</u>, white, yellow, grey,
green, red, blue; Hardness 8;
Locality: Minas Gerais, Brasil

Photography by Jim Hughes, Assisted by ♪ Melody ♫
Collection of Bob Jackson, Applewood, CO, USA

TOPAZ - (Sheet) -
Al₂SiO₄(F,OH)₂ - Translucent
colourless, white, yellow, grey,
green, red, blue; Hardness 8;
Locality: Minas Gerais, Brasil

Photography by Jim Hughes, Assisted by ♪ Melody ♫
Collection of ♪ Melody ♫, Applewood, CO, USA

TOPAZ - (Sheet) -
Al₂SiO₄(F,OH)₂ - Translucent
colourless, white, yellow, grey,
green, red, blue; Hardness 8;
Locality: Minas Gerais, Brasil

Photography by Jim Hughes, Assisted by ♪ Melody ♫
Collection of ♪ Melody ♫, Applewood, CO, USA

TOPAZ - (Tri-Coloured) -
$Al_2SiO_4(F,OH)_2$ - Translucent
colourless, white, yellow, grey,
green, red, blue; Hardness 8;
Locality: Minas Gerais, Brasil

Photography by Jim Hughes, Assisted by ♪ Melody ♫
Collection of ♪ Melody ♫, Applewood, CO, USA

TOUCHSTONE - SiO_2 with
impurities; Velvety-black siliceous
stone; Hardness 7; Locality:
Oregon, USA

Photography by Jim Hughes, Assisted by ♪ Melody ♫
Collection of ♪ Melody ♫, Applewood, CO, USA

TOURMALINE - Green with
Quartz) - (Na,K,Ca)
$(Mg,Fe,Mn,Li,Al)_3(Al,Fe,Cr,V)_6Si_6$
$O_{18}(BO_3)_3(O,OH,F)_4$ - [General
ring silicate]; Hardness 7-7.5;
Locality: Minas Gerais, Brasil

Photography by Jim Hughes, Assisted by ♪ Melody ♫
Collection of Bob Jackson, Applewood, CO, USA

TOURMALINE - *(Purple, Light Pink, Green, Blue, Rubellite, etc.)* - *(Na,K,Ca)(Mg,Fe,Mn,Li,Al)$_3$ (Al,Fe,Cr,V)$_6$Si$_6$O$_{18}$(BO$_3$)$_3$ (O,OH,F)$_4$* - *[General ring silicate]; Hardness 7-7.5; Locality: Minas Gerais, Brasil*

Photography by Jim Hughes, Assisted by ♩ Melody ♫
Collection of ♩ Melody ♫, Applewood, CO, USA

TOURMALINE - *(Blue with Quartz)* - *(Na,K,Ca) (Mg,Fe,Mn,Li,Al)$_3$(Al,Fe,Cr,V)$_6$ Si$_6$O$_{18}$(BO$_3$)$_3$(O,OH,F)$_4$* - *[General ring silicate]; Hardness 7-7.5; Locality: Minas Gerais, Brasil*

Photography by Jim Hughes, Assisted by ♩ Melody ♫
Collection of ♩ Melody ♫, Applewood, CO, USA

TOURMALINE - *(Black)* - *(Na,K,Ca)(Mg,Fe,Mn,Li,Al)$_3$ (Al,Fe,Cr,V)$_6$Si$_6$O$_{18}$(BO$_3$)$_3$ (O,OH,F)$_4$* - *[General ring silicate]; Hardness 7-7.5; Specifically NaFe$_3$Al$_6$ (BO$_3$)$_3$Si$_6$O$_{18}$(OH)$_4$; Locality; Minas Gerais, Brasil*

Photography by Jim Hughes, Assisted by ♩ Melody ♫
Collection of ♩ Melody ♫, Applewood, CO, USA

TOURMALINE - *(Watermelon With Paraiba)* - *(Na K,Ca)* *(Mg,Fe,Mn,Li,Al)$_3$(Al,Fe,Cr,V)$_6$ Si$_6$O$_{18}$(BO$_3$)$_3$(O,OH,F)$_4$ - [General ring silicate]; Hardness 7-7.5; Locality: Minas Gerais, Brasil*

Photography by Jim Hughes, Assisted by ♪ Melody ♫
Collection of ♪ Melody ♫, Applewood, CO, USA

TOURMALINE - *(Watermelon)* - *(Na,K,Ca)(Mg,Fe,Mn,Li,Al)$_3$ (Al,Fe,Cr,V)$_6$Si$_6$O$_{18}$(BO$_3$)$_3$ (O,OH,F)$_4$ - [General ring silicate]; Hardness 7-7.5; Locality: Minas Gerais, Brasil*

Photography by ♪ Melody ♫
Collection of Bob Jackson, Applewood, CO, USA

TOURMALINE - *(Watermelon With Neon Blue)* - *(Na K,Ca)* *(Mg,Fe,Mn,Li,Al)$_3$(Al,Fe,Cr,V)$_6$ Si$_6$O$_{18}$(BO$_3$)$_3$(O,OH,F)$_4$ - [General ring silicate]; Hardness 7-7.5; Locality: Africa*

Photography by Jim Hughes, Assisted by ♪ Melody ♫
Collection of ♪ Melody ♫, Applewood, CO, USA

TOURMALINE - *(Watermelon With Black Center)* - *(Na K,Ca) (Mg,Fe,Mn,Li,Al)$_3$(Al,Fe,Cr,V)$_6$ Si$_6$O$_{18}$(BO$_3$)$_3$(O,OH,F)$_4$ - [General ring silicate]; Hardness 7-7.5; Locality: Minas Gerais, Brasil*

Photography by Jim Hughes, Assisted by ♪ Melody ♫
Collection of ♪ Melody ♫, Applewood, CO, USA

TOURMALINE - *(Pink and Rubellite)* - *(Na K,Ca) (Mg,Fe,Mn,Li,Al)$_3$(Al,Fe,Cr,V)$_6$ Si$_6$O$_{18}$(BO$_3$)$_3$(O,OH,F)$_4$ - [General ring silicate]; Hardness 7-7.5; Locality: California, USA*

Photography by Jim Hughes, Assisted by ♪ Melody ♫
Collection of ♪ Melody ♫, Applewood, CO, USA

TOURMALINE - *(Blackish-Brown)* - *(Na K,Ca) (Mg,Fe,Mn,Li,Al)$_3$(Al,Fe,Cr,V)$_6$ Si$_6$O$_{18}$(BO$_3$)$_3$(O,OH,F)$_4$ - [General ring silicate]; Hardness 7-7.5; Locality: West Australia, Australia*

Photography by Jim Hughes, Assisted by ♪ Melody ♫
Collection of ♪ Melody ♫, Applewood, CO, USA

TOURMALINE - (Purple/Blue) - (Na K,Ca) (Mg,Fe,Mn,Li,Al)$_3$ (Al,Fe,Cr,V)$_6$Si$_6$O$_{18}$(BO$_3$)$_3$ (O,OH,F)$_4$ - [General ring silicate]; Hardness 7-7.5; Locality: Minas Gerais, Brasil

Photography by Jim Hughes, Assisted by ♪ Melody ♫
Collection of ♪ Melody ♫, Applewood, CO, USA
Gift from Bob Jackson, Applewood, CO, USA

TOURMALINE - (Orange, Green, Pink) - (Na K,Ca) (Mg,Fe,Mn,Li,Al)$_3$(Al,Fe,Cr,V)$_6$Si$_6$O$_{18}$(BO$_3$)$_3$(O,OH,F)$_4$ - [General ring silicate]; Hardness 7-7.5; Locality: Minas Gerais, Brasil

Photography by Jim Hughes, Assisted by ♪ Melody ♫
Collection of ♪ Melody ♫, Applewood, CO, USA

TOURMALINE - (Purple, Yellow, Green) - (Na K,Ca) (Mg,Fe,Mn,Li,Al)$_3$(Al,Fe,Cr,V)$_6$Si$_6$O$_{18}$(BO$_3$)$_3$(O,OH,F)$_4$ - [General ring silicate]; Hardness 7-7.5; Locality: Minas Gerais, Brasil

Photography by Jim Hughes, Assisted by ♪ Melody ♫
Collection of ♪ Melody ♫, Applewood, CO, USA

TOURMALINE - (Orange, Yellow, Etc.) - (Na K,Ca) (Mg,Fe,Mn,Li,Al)$_3$(Al,Fe,Cr,V)$_6$ Si$_6$O$_{18}$(BO$_3$)$_3$(O,OH,F)$_4$ - [General ring silicate]; Hardness 7-7.5; Locality: Minas Gerais, Brasil

Photography by Jim Hughes, Assisted by ♪ Melody ♫
Collection of ♪ Melody ♫, Applewood, CO, USA

TOURMALINE - (Orange, Green, Purple) - (Na K,Ca) (Mg,Fe,Mn,Li,Al)$_3$(Al,Fe,Cr,V)$_6$ Si$_6$O$_{18}$(BO$_3$)$_3$(O,OH,F)$_4$ - [General ring silicate]; Hardness 7-7.5; Locality: Minas Gerais, Brasil

Photography by Jim Hughes, Assisted by ♪ Melody ♫
Collection of ♪ Melody ♫, Applewood, CO, USA

TOURMALINE - (Paraiba) - (Na K,Ca) (Mg,Fe,Mn,Li,Al)$_3$ (Al,Fe,Cr,V)$_6$Si$_6$O$_{18}$(BO$_3$)$_3$(O,O H,F)$_4$ - [General ring silicate]; Hardness 7-7.5; Locality: Minas Gerais, Brasil

Photography by Jim Hughes, Assisted by ♪ Melody ♫
Collection of ♪ Melody ♫, Applewood, CO, USA

TOURMALINE - *(Rubellite [With Lepidoite]) - (Na K,Ca) $(Mg,Fe,Mn,Li,Al)_3(Al,Fe,Cr,V)_6Si_6O_{18}(BO_3)_3(O,OH,F)_4$ - General ring silicate]; Hardness 7-7.5; Locality: California, USA*

Photography by Jim Hughes, Assisted by ♪ Melody ♫
Collection of Bob Jackson, Applewood, CO, USA

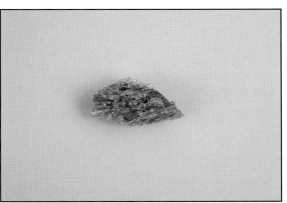

TREMOLITE - *$Ca(Mg,Fe)_5Si_8O_{22}(OH)_2$ - Vitreous colourless, grey; Hardness 5-6; Locality: Ontario, Canada*

Photography by Jim Hughes, Assisted by ♪ Melody ♫
Collection of ♪ Melody ♫, Applewood, CO, USA

TREVORITE - *(With Willemeseite) - $NiFe_2O_4$ - Metallic/sub-metallic brown; Hardness 5; Locality: Transvaal, Republic of South Africa, Africa*

Photography by Jim Hughes, Assisted by ♪ Melody ♫
Collection of ♪ Melody ♫, Applewood, CO, USA

TRILOBITE - *Fossilized sea creature from eons past; Locality: British Columbia, Canada*

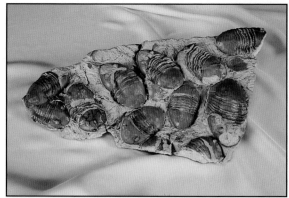

Photography by Jim Hughes, Assisted by ♪ Melody ♫
Collection of Colorado School of Mines, CO, USA

TRIPHYLITE - *LiFePO₄ - Vitreous/sub-resinous bluish-grey, greenish-grey; Hardness from 4-5; Locality: New Hampshire, USA*

Photography by Jim Hughes, Assisted by ♪ Melody ♫
Collection of ♪ Melody ♫, Applewood, CO, USA

TRIPLOIDITE - *(In Lithiophilite) (Mn,Fe)₂PO₄(OH) - Vitreous/lustrous/adamantine <u>pinkish</u> to deep rose, yellow, yellowish-brown; Hardness from 4.5-5; Locality: Rio Grande Do Norte, Brasil*

Photography by Dave Shrum, Colorado Camera Co., Lakewood, Colorado, USA
Collection of ♪ Melody ♫, Applewood, CO, USA

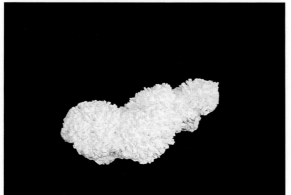

TRONA - $Na_3(HCO_3)(CO_3)$♥$2H_2O$ - Vitreous, colourless, grey, yellowish-white, <u>white</u>; Hardness 2.5-3; Locality: Utah, USA

Photography by Jim Hughes, Assisted by ♪ Melody ♫
Collection of Bob Jackson & ♪ Melody ♫, Applewood,
CO, USA; Gift of Dallas Noyes, Richland, WA, USA

TSAVORITE - (Crystal) - $Ca_3Al_2Si_3O_{12}$ (With Vanadium); Translucent to opaque vitreous emerald green; Hardness 6.5-7; Locality: Kenya, Africa

Photography by Jim Hughes, Assisted by ♪ Melody ♫
Collection of ♪ Melody ♫, Applewood, CO, USA

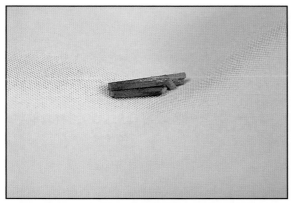

TSUMEBITE - $CuPb_2(PO_4)(SO_4)(OH)$ - Vitreous green; Hardness 3.5; Locality: Tsumeb, Namibia

Photography by Jim Hughes, Assisted by ♪ Melody ♫
Collection of ♪ Melody ♫, Applewood, CO, USA

TUFA - $CaCO_3$ - *Porous formation of calcium carbonate; Hardness 2-3; Locality: New Mexico, USA*

Photography by Jim Hughes, Assisted by ♪ Melody ♫
Collection of Bob Jackson, Applewood, CO, USA

TUGTUPITE - *(In Nepheline Syenite)* - $Na_4BeAlSi_4O_{12}Cl$ - *Translucent white, <u>pink to deep red</u>, pale blue; Hardness 5.5-6.5; Locality: Greenland*

Photography by Jim Hughes, Assisted by ♪ Melody ♫
Collection of ♪ Melody ♫, Applewood, CO, USA

TUNNELITE - *Composition unknown to author; Locality: Kern County, California, USA*

Photography by Jim Hughes, Assisted by ♪ Melody ♫
Collection of ♪ Melody ♫, Applewood, CO, USA

TUNGSTENITE - WS_2 - *Metallic grey, white; Hardness 2.5; Locality: Utah, USA*

Photography by Jim Hughes, Assisted by ♪ Melody ♫
Collection of Colorado School Of Mines, CO, USA

TURGITE - $2Fe_2O_3$♥H_2O - *Crimson to brown in thin fibers; Hardness: 6.5; Locality: Nevada, USA*

Photography by Jim Hughes, Assisted by ♪ Melody ♫
Collection of ♪ Melody ♫, Applewood, CO, USA

TURQUOISE - *(Chinese)* - $CuAl_6(PO_4)_4(OH)_8$♥$4H_2O$ - *Vitreous blue, bluish-green, green; Hardness 5-6; Locality: China*

Photography by Jim Hughes, Assisted by ♪ Melody ♫
Collection of Bob Jackson, Applewood, CO, USA

Take care of your stones,
and they'll take care of you.

[Barbara Manojlvich,
World of Gems, North Carolina, USA]

ULEXITE - NaCaB₅O₆(OH)₆♥5H₂O - Silky, vitreous colourless, white; Hardness 2.5;
Location: California, USA; Photography by Jim Hughes; Assisted by ♪ Melody ♫;
Collection of ♪ Melody ♫, Applewood, CO, USA

ULEXITE - NaCaB₅O₆(OH)₆♥5H₂O - Silky, vitreous colourless, white; Hardness 2.5;
Location: California, USA; Photography by Jim Hughes; Assisted by ♪ Melody ♫;
Collection of ♪ Melody ♫, Applewood, CO, USA

ULLMANNITE - *NiSbS* - *Metallic grey; Hardness 5-5.5; Locality: Czechoslavakia*

Photography by Jim Hughes; Assisted by ♪ Melody ♫
Collection of ♪ Melody ♫, Applewood, CO, USA

UNAKITE - *Pink feldspar, green epidote, and quartz; Hardness 6-7; Locality: Virginia, USA*

Photography by Jim Hughes, Assisted by ♪ Melody ♫
Collection of ♪ Melody ♫, Applewood, CO, USA

URANINITE - *A Uranate of Uranyl, lead, thorium, lanthanum, and yttrium, with calcium and water; Hardness approximately 5.5; Locality: North West Territory, Canada*

Photography by Jim Hughes, Assisted by ♪ Melody ♫
Collection of ♪ Melody ♫, Applewood, CO, USA

UVAROVITE - Ca₃Cr₂(SiO₄)₃ -
Vitreous/resinous green;
Hardness 7.5; Locality:
California, USA

Photography by Jim Hughes, Assisted by ♪ Melody ♫
Collection of ♪ Melody ♫, Applewood, CO, USA

UVAROVITE - Ca₃Cr₂(SiO₄)₃ -
Vitreous/resinous green;
Hardness 7.5; Locality: Ural
Mts., Russia

Photography by Jim Hughes, Assisted by ♪ Melody ♫
Collection of ♪ Melody ♫, Applewood, CO, USA

VALENCIANITE - $KAlSi_3O_8$ -
An adularia found only in the
silver mine of Valencia, Mexico;
Hardness 6.

Photography by Jim Hughes, Assisted by ♪ Melody ♫
Collection of ♪ Melody ♫, Applewood, CO, USA

VALENTINITE - Sb_2O_3 -
Adamantine colourless, white,
yellowish, reddish; Hardness
from 2.5 - 3; Locality: Quebec,
Canada

Photography by Jim Hughes, Assisted by ♪ Melody ♫
Collection of ♪ Melody ♫, Applewood, CO, USA

VANADINITE - $Pb_5(VO_4)_3Cl$;
Sub-resinous/sub-adamantine
orange, red, brown, yellow, etc.;
Hardness 2.7-3; Locality:
Arizona

Photography by Jim Hughes, Assisted by ♪ Melody ♫
Collection of Bob Jackson, Applewood, CO, USA

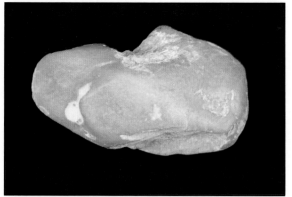

VARISCITE - $AlPO_4 \heartsuit 2H_2O$ -
Vitreous pale/emerald green;
Hardness 3.5-4.5; Locality:
Arizona, USA

VARISCITE - *(In Matrix)* -
$AlPO_4 \heartsuit 2H_2O$ - *Vitreous*
pale/emerald green; Hardness 3.5-
4.5; Locality: Arizona, USA

VAUXITE -
$FeAl_2(PO_4)_2(OH)_2 \heartsuit 6H_2O$ -
Vitreous blue; Hardness 3.5;
Locality: Bolivia

VERDITE - $H_4Mg_3Si_2O_9$ with dolomite, magnesite or calcite and sometimes veined with <u>white</u> or pale green and/or mottled with red or yellow; Clouded green colour with possible white/green veining, sub-resinous and translucent to opaque; Hardness 2.5-4; The Talc variety is shown; Locality: Maryland, USA

Photography by Jim Hughes, Assisted by ♪ Melody ♫
Collection of ♪ Melody ♫, Applewood, CO, USA

VESUVIANITE - $Ca_{19}Fe(Mg,Al)_8Al_4(SiO_4)_{10}(Si_2O_7)_4(OH)_{10}$ - *Vitreous/resinous brown, green, yellow, pale blue; Hardness 6.5; Locality: Maine, USA*

Photography by Jim Hughes, Assisted by ♪ Melody ♫
Collection of ♪ Melody ♫, Applewood, CO, USA

VESZELYITE - $(Cu,Zn)_3PO_4(OH)_3$♥$2H_2O$ - *Vitreous greenish-blue, dark blue; Hardness 3.5-4; Locality: Montana, USA*

Photography by Jim Hughes, Assisted by ♪ Melody ♫
Collection of ♪ Melody ♫, Applewood, CO, USA

***VIVIANITE** (Crystal)-* $Fe_3P_2O_8$ *- Pearly luster, vitreous colourless, blue to green, light to dark; Hardness 1.5-2; Locality: Bolivia. Photography by Jim Hughes, Assisted by ♪ Melody ♫; Collection of ♪ Melody ♫, Applewood, CO, USA*

***VIVIANITE** (Crystal)-* $Fe_3P_2O_8$ *- Pearly luster, vitreous colourless, blue to green, light to dark; Hardness 1.5-2; Locality: Ukraine. Photography by Jim Hughes, Assisted by ♪ Melody ♫; Collection of ♪ Melody ♫, Applewood, CO, USA*

Begin by living each moment as if you had faith ...
and you will ♪

WAD - A Manganese oxide in amorphous and reniform masses, either earthy or compact; Usually very soft, and less often with hardness of 6; Range of colour includes blacks, blues, browns, etc.; Location: Arizona, USA

*Photography by Jim Hughes, Assisted by ♪ Melody ♫
Collection of ♪ Melody ♫, Applewood, CO, USA*

WADEITE - (In Microcline) - $K_2ZrSi_3O_9$ -Translucent colourless to pink; Hardness 5-6; Locality: Kola Peninsula, Russia

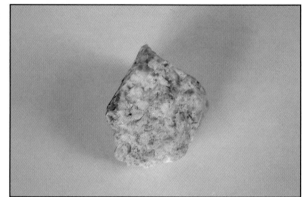

*Photography by Jim Hughes, Assisted by ♪ Melody ♫
Collection of ♪ Melody ♫, Applewood, CO, USA*

WAGNERITE - $(Mg,Fe)_2PO_4F$ - Vitreous yellow, greyish, <u>pink</u> to red, greenish; Hardness 5-5.5; Locality: Norway

*Photography by Jim Hughes, Assisted by ♪ Melody ♫
Collection of ♪ Melody ♫, Applewood, CO, USA*

WAKEFIELDITE - $(Ce,Pb)VO_4$ - Black (in vug) yellow in light; Hardness 4.5; Locality: British Columbia, Canada

Photography by Jim Hughes, Assisted by ♪ Melody ♫
Collection of ♪ Melody ♫, Applewood, CO, USA

WARDITE - $NaAl_3(PO_4)_2$ $(OH)_4♥2H_2O$ - Vitreous blue-green, pale green, colourless; Hardness 5; Locality: Yukon Territory, Canada

Photography by Jim Hughes, Assisted by ♪ Melody ♫
Collection of ♪ Melody ♫, Applewood, CO, USA

WAVELLITE - $Al_3(PO_4)_2$ $(OH,F)_3♥5H_2O$ - Vitreous/pearly/resinous green, yellow, white, brown, etc.; Hardness 3.2-4; Locality: California, USA

Photography by Jim Hughes, Assisted by ♪ Melody ♫
Collection of ♪ Melody ♫, Applewood, CO, USA

WAVELLITE - $Al_3(PO_4)_2(OH,F)_3 \heartsuit 5H_2O$ - *Vitreous/<u>pearly</u>/resinous green, yellow,*
<u>white</u>, *brown, etc.; Hardness 3.2-4; Locality: California; Photography by*
Jim Hughes, Assisted by ♪ Melody ♫; Collection of ♪ Melody ♫, Applewood, CO, USA

WEEKSITE (With Yellow Carnotite) - Composition unknown to author; Locality:
Arkansas, USA; Photography by Jim Hughes, Assisted by ♪ Melody ♫
Collection of ♪ Melody ♫, Applewood, CO, USA

WHERRYITE -
$CuPb_4O(SO_4)_2(CO_3)(OH,Cl)_2$ -
Translucent pale yellow,
yellowish-green to blue, light
green; Hardness ?; Locality:
Arizona, USA

Photography by Jim Hughes, Assisted by ♪ Melody ♫
Collection of ♪ Melody ♫, Applewood, CO, USA

WHITLOCKITE (Crystals)-
$Ca_{18}(Mg,Fe)_2(H_2,Ca)(PO_4)_{14}$ -
Vitreous/sub-resinous colourless,
<u>pinkish</u>, white, grey, yellowish;
Hardness 5; Locality: South
Dakota, USA

Photography by Jim Hughes, Assisted by ♪ Melody ♫
Collection of ♪ Melody ♫, Applewood, CO, USA

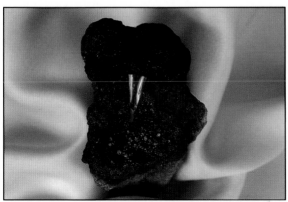

WHITLOCKITE (Crystals)-
$Ca_{18}(Mg,Fe)_2(H_2,Ca)(PO_4)_{14}$ -
Vitreous/sub-resinous colourless,
pinkish, <u>white</u>, grey, yellowish;
Hardness 5; Locality: New
Hampshire, USA

Photography by Jim Hughes, Assisted by ♪ Melody ♫
Collection of ♪ Melody ♫, Applewood, CO, USA

WILKEITE -
$3Ca_3(PO_4)_2$♥$CaCO_3$♥$3Ca_3[(SiO_4)(SO_4)]$♥CaO - *Vitreous to sub-resinous pale pink and yellow; Hardness 5; Locality: Mexico*

Photography by Jim Hughes, Assisted by ♪ Melody ♫
Collection of Colorado School Of Mines, CO, USA

WILLEMITE - Zn_2SiO_4 -
Vitreo-resinous white, greenish-yellow, green, red, etc.; Hardness 5.5; Locality: New Jersey, USA

Photography by Jim Hughes, Assisted by ♪ Melody ♫
Collection of ♪ Melody ♫, Applewood, CO, USA

WILLIAMSITE - $H_4Mg_3Si_2O_9$ -
Sub-resinous to earthy, <u>green</u>, red, yellow; Hardness 2.5-4; Locality: North Carolina, USA

Photography by Jim Hughes, Assisted by ♪ Melody ♫
Collection of ♪ Melody ♫, Applewood, CO, USA

WITHERITE - $BaCO_3$ -
Vitreous/resinous <u>colourless,</u>
white, greyish; Hardness 3.3-5;
Locality: Illinois, USA

WOHLERITE - *(Crystals in*
Zircon/Syenite) -
$Na_2Ca_4ZrNb(Si_2O_7)_2(O,F)_4$ -
Vitreous/resinous pale yellowish,
white, brownish, greyish;
Hardness 5-6; Locality: Norway

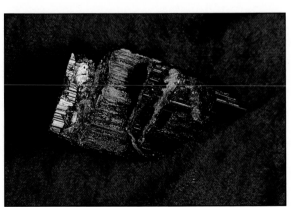

WOLFRAMITE - $(Fe,Mn)WO_4$ -
Sub-metallic/adamantine greyish-
black; Hardness 4-4.5; Locality:
Portugal

WOODHOUSEITE -
$CaAl_3(SO_4)(PO_4)(OH)_6$ -
Vitreous colourless, flesh to yellow coloured, white; Hardness 4.5; Locality: California, USA

Photography by Jim Hughes, Assisted by ♪ Melody ♫
Collection of ♪ Melody ♫, Applewood, CO, USA

WOODWARDITE -
$(Cu,Al)_8SO_4(OH)_{16}♥nH_2O$ -
Translucent greenish-blue, turquoise-blue; Hardness ?; Locality: Ireland

Photography by Jim Hughes, Assisted by ♪ Melody ♫
Collection of ♪ Melody ♫, Applewood, CO, USA

WULFENITE - $PbMoO_4$ -
Resinous/adamantine orange-yellow, yellow, grey, brown, etc.; Hardness 2.7-3; Locality: Sonora, Mexico

Photography by Jim Hughes, Assisted by ♪ Melody ♫
Collection of Julianne Guilbault, Lakewood, CO, USA

THE way is always open to you ♫

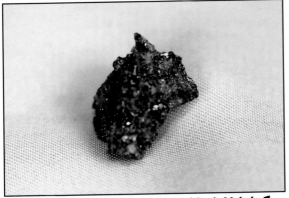

XANTHOCONITE - Ag_3AsS_3 - *Adamantine red/orange/clove-brown; Hardness 2-3; Peru, South America*

Photography by Jim Hughes, Assisted by ♪ Melody ♫
Collection of ♪ Melody ♫, Applewood, CO, USA

XENOTIME - YPO_4 - *Vitreous/resinous yellowish/reddish brown, yellow, red, etc.; Hardness 4-5; Locality: Bahia, Brasil*

Photography by Jim Hughes, Assisted by ♪ Melody ♫
Collection of ♪ Melody ♫, Applewood, CO, USA

XONOTLITE - $Ca_6Si_6O_{17}(OH)_2$ - *Translucent white, grey, pale pink; Hardness 6.5; Locality: Arizona, USA.*

Photography by Jim Hughes, Assisted by ♪ Melody ♫
Collection of ♪ Melody ♫, Applewood, CO, USA

The way of peace is to become oneself ♫